# Writing a Built Environment Dissertation

# Writing a Built Environment Dissertation
## Practical guidance and examples

**Dr Peter Farrell MSc FRICS FCIOB FRSA**
Reader and Programme Leader
MSc in Construction Management
University of Bolton
UK

A John Wiley & Sons, Ltd., Publication

*Library of Congress Cataloging-in-Publication Data*

Farrell, Peter, 1955-
   Writing a built environment dissertation : practical guidance and examples/Peter Farrell.
       p. cm.
   Includes bibliographical references and index.
   ISBN 978-1-4051-9851-6 (alk. paper)
   1. Building–Research. 2. Technical writing. 3. Academic writing. 4. Dissertations, Academic–Authorship. I. Title.
   TH213.5.F37 2011
   808'.06669–dc22

                                                            2010029196

A catalogue record for this book is available from the British Library.

This book is published in the following electronic formats: ePDF [9781444328677]

Set in 10/12pt Minion by Thomson Digital, Noida, India.
Printed and bound in Malaysia by Vivar Printing Sdn Bhd

2   2012

10 0670104 2

# Contents

# Author Biography

Peter Farrell lectures at the University of Bolton, and has taught research methods to under and postgraduate students for over 10 years. He also delivers modules and undertakes research in construction management and commercial management. His industry training was in planning and quantity surveying, whilst most of his post-qualification work was as a contractor's site manager.

# Preface

The aim of this book is to provide practical guidance on the preparation of undergraduate dissertations in the built environment. It is hoped that it will give students the platform to attain their maximum possible mark. Some sections of the book may contribute towards enhanced performance in other modules. For example, suggestions about how to develop theory and use literature as part of a critical appraisal are common to many subjects in the built environment and indeed other disciplines. The book is ordered around a structure that may be useful for a dissertation; that is, it starts with what material should be contained in an introduction chapter and finishes with material that should be in the conclusion. Embedded throughout the book are discussions of issues around study skills and ethics. There are many examples included, using a variety of methodological designs, in which students are encouraged to consider the concepts of reliability and validity. A key difference between a dissertation and other courseworks is that the middle of the dissertation should include a data collection process and some analysis. Suggestions are made about how to collect data and how to perform analysis. The analytical chapters cover qualitative and quantitative approaches. The qualitative chapter demonstrates how to include some rigour in the analytical process, as students often rely on simplistic browsing of material. The quantitative chapter attempts to avoid some of the complexity in statistical work without devaluing its usefulness. The book encourages students to undertake a process of self reflection at the end of their research, and to include a section on limitations and criticisms of their own studies. It is hoped that the examples used will stimulate ideas about how students can develop their chosen topic area into dissertation format.

# 1 Introduction

The objectives of each section of this chapter are:

1.1 Introduction: to set the scene and describe the dissertation process
1.2 Terminology and nomenclature: to emphasise the importance of the objective
1.3 Document structure: to provide a template
1.4 Possible subject areas for your dissertation: suggest topic areas and encourage early reading
1.5 Qualitative and quantitative analysis: to distinguish between the two analytical schools
1.6 The student/supervisor relationship and time management: to provide templates
1.7 Ethical codes: to identify ground rules for compliance with codes of practice
1.8 House style or style guide: to promote consistency and provide a template
1.9 Writing style: to identify potential pitfalls
1.10 Proofreading: to encourage it as a process, using if necessary independent help

## 1.1 Introduction

In some universities the dissertation may carry as much as one-quarter weighting towards the final year degree classification. It is the flagship document of your study. It is the document that external examiners will look at with greatest scrutiny. You may want to take it to your employer and/or prospective employers. You will hopefully be proud to show it to members of your family, and it will sit on your bookshelf so that you can show it to your grandchildren. It is a once in a lifetime journey for most; it is to be enjoyed, and remembered.

One of the key criteria for a dissertation is that it must have some originality. That is, it need not discover something new, but perhaps merely look at an area that has already been investigated, and take a different perspective on it or use a different methodology. It is more than an assignment – the research process must seek the information, analyse it and offer conclusions. Modest objectives are adequate. Better dissertations have robust methods of analysing qualitative data or some basic statistical analysis.

*Writing a Built Environment Dissertation: Practical Guidance and Examples.* Peter Farrell.
©2011 Peter Farrell. Published 2011 by Blackwell Publishing Ltd.

Dissertations have assessment criteria. To achieve marks in the upper echelons (70% plus), the criteria often require that work should demonstrate 'substantial evidence of originality and creativity', 'very effective integration of theory and practice', 'excellent grasp of theoretical, conceptual, analytical and practical elements', and 'all information/skills deployed'.

There are two separate strands to your dissertation. The first is that you must develop your knowledge in your chosen topic so that you become 'expert'. One of the reasons you may have chosen your subject is that you may want to learn more about it. It is indeed very important that you do this. The second strand is that you must conduct a piece of research, employing appropriate research methodology. In your document you must explain and substantiate your methodology; it must stand up to scrutiny. The method that you use must include the collection and analysis of data. The two strands go hand-in-hand. It is not to say that the weighting is 50:50, or any other percentage, but there must be substantial evidence of both strands in your dissertation.

## 1.2 Terminology; nomenclature

Clarity in research is absolutely critical; the plethora of terminology used by academics can be unhelpful, fuzzy, and for some misleading. That is just the way it is. It may be useful for you to employ your own rigid definitions of such terminology, or at the very least be consistent, in the language you use in your work. For example, in describing what a research project will 'do' students often express this by using different words in the same document. An often used and favoured term by researchers is 'objective'. Frequently students stray from this word by stating initially 'the objective of this study is . . .', but elsewhere it becomes 'the focus of the study', 'the reason for the study', 'the study looks into', 'the study tries to', 'the study examines', 'purpose', 'goal', 'direction', 'intention', or 'seeks to'. However, whilst being rigid (or more appropriately consistent) within your own work, it must be recognised that universities and individual academics will have their own preferences, and students must be able to adapt flexibly to work with supervisors, and also to understand the writing of others who use different language. Most supervisors will be comfortable that you 'hang' the whole of your study around objectives; put more clearly, objectives, objectives and objectives.

## 1.3 Document structure

A suggested structure/template for a dissertation is:

- No number: Preliminary pages
- Chapter 1: Introduction
- Chapter 2: Theory and literature review
- Chapter 3: Research design and methodology
- Chapter 4: Analysis, results and findings
- Chapter 5: Discussion
- Chapter 6: Conclusions and recommendations
- No number: References and bibliography
- No number: Appendices

This is not written in tablets of stone, but is merely a framework around which your structure may be designed. It is for individual researchers to design their structure, and to agree it with their supervisor. The suggestions may be considered as chapter titles, but they should be 'flavoured' by words relevant to your study area, e.g. 'the development of theory and literature about money as a motivator for construction craftspeople'.

Each chapter or number of words do not necessarily lend themselves to one-sixth weight. There is an argument that the first two chapters, as the opening to the dissertation, could be about one-third weight. The middle two chapters, comprising the analytical framework, could be about one-third weight and finally the last two chapters, closing off the dissertation, about one-third weight. Often it is the latter part where students lose marks; they simply run out of time after completing the analysis. The consequence is that dissertations destined for really good marks only achieve mid-range marks.

Each chapter should open with an introduction – even an introduction to the introduction chapter - and close with a summary. Students often do not like writing either introductions or summaries, and question their value to readers. The introduction to each chapter need only be a few paragraphs. It is not for readers to embark on a voyage of discovery as they read each chapter. The introduction chapter should set the scene. The 'introduction to the introduction' may start with the aim of the study. It may tell the reader that the introduction chapter will: provide a background to the topic area and description of the problem; give a historical perspective; give the research goals (including the objectives); describe briefly the methodology; give an outline of the remaining parts of the dissertation; and summarise the chapter. However, it should not be written as mechanically as stated above. It should be flavoured by your topic area, e.g. an historical perspective of private finance initiative (PFI) as a procurement method. The writing style of a summary is different to the writing style of an introduction. It does exactly what its name implies; it summarises what has gone before. It should not say 'this chapter has outlined the problem'. It should in the narrative summarise the key points of the problem in a few lines. You need to say what the problem is. A useful tactic when writing a summary is to read each page and condense it into one or two carefully selected sentences. The reason for a summary is that readers, who have taken the journey through your chapter, may need some moments of thought and reflection about what they have just read, before going on. They may indeed have forgotten what they read at the front of the chapter by the time they get to the end. Also, readers may not read the whole document in one sitting. When they come to recommence reading, the summary can refresh minds before continuing.

The whole document should be in report numbering format. Start with the introduction chapter as Chapter 1. The introduction to the introduction is 1.1, 1.2 is definitions of important phrases, 1.3 is background to the topic area, etc. Try to avoid too many sub-sub-sections, but if they are needed they become for example 1.3.1, 1.3.2, etc.

Page number the whole document, except the cover page. By convention, preliminary pages are numbered with Roman numerals, that is (i), (ii), etc. The first page is a declaration, numbered Roman numeral (i). People with dyslexia may find it hard to distinguish between Roman numerals; therefore alternatively consider letters (a), (b), (c), etc. Pages after the preliminary pages, starting with the cover page to Chapter 1, use Arabic numerals 1, 2, 3, etc. The cover page to Chapter 1 thus starts at page 1. Page numbering with Arabic numerals continues into the reference section and the appendices. Separate parts of the appendices are labelled by letters; that is appendix 'A' may be a covering letter to a postal questionnaire, appendix 'B' may be the questionnaire itself, etc.

The preliminary pages to a research document should include the following separate parts:

(a) Unnumbered - a cover page with the document title, name of author, name of university, year and degree title.

(b) Declaration using words prescribed by the university such as 'I declare that this research has not been submitted to any other university or institution of learning'.

(c) Acknowledgements page - it is usually to thank people who have contributed to the research through time or sponsorship, employers, friends or members of your family and supervisors. Only a short statement is usual.

(d) Abstract - the abstract is a very concise summary, and should be written carefully. Readers may be initially attracted to documents by titles, but these can be misleading, and more information is required. Its purpose is to allow readers to make a quick decision about whether they wish to read further sections of the document, or alternatively they may be able to make a sensible judgement that the document is not relevant to their needs. Often readers who are browsing previous research will read abstracts and decide not to read on; that is fine. They have been able to make an informed decision quickly based upon a full concise summary of the document. Since you have a limited amount of words, and you may wish to entice people into the document, each part must be measured carefully. External examiners will read some, but cannot read all dissertations. Given a choice of documents, they may be attracted to read documents by a well articulated abstract. In academic publications abstracts are often 200 to 250 words in length, but in dissertations perhaps a larger word limit is acceptable. An abstract confined neatly to one A4 page of text, single line space, 12 size font, perhaps three or four paragraphs with line spaces, would comprise about 500 words. Try to avoid going onto a second page, even for one line. This is your opportunity to sell your work. In research terms, it would be a serious failing if subsequent researchers picked up your document, with the idea to further knowledge in your field, but because of a lack of clarity in the abstract, were led to think your work was not relevant. If a sentence, or indeed a single word is not necessary to convey the message required, it should be taken out. The abstract is an art in writing concisely and with precision. It should: give the topic, state the aim, outline the problem, give the main objectives or hypotheses, summarise the methodology (including population description, sample size if appropriate, method of data collection and analyses), state the main findings, conclusions and recommendations. It can be written as work proceeds but can only be completed at the end. Students often adopt a writing style for an abstract similar to the following: 'the study will give an objective, and describe the methodology...', etc. This is not an abstract, since it would leave readers without the information required. The abstract must actually state what the objective is, and state the methodology. Some students submit their documents without an abstract; deduct five marks!

(e) Contents page – this may list the main title of each chapter. It is not usually necessary to list all sub-sections of chapters on the main contents page. Subsequently, each chapter should have its own cover page that details the titles to sub-sections within the chapter.

(f) List of abbreviations.

(g) Glossary of symbols, if statistical tests are executed. Letters of the Greek alphabet are often used to distinguish between different tests.

(h)    Glossary of terms - this ensures a common understanding even for quite well-known terms. Also terms which have a particular meaning in the subject topic of the research can be explained. It will include a brief definition of terms in the context of the study. Ensure that such definitions are authoritative; that is, from the literature.
(i)    List of appendices.
(j)    List of figures.
(k)    List of tables.

Tables may contain results of experiments, or summarise data. Figures may be pie charts, histograms, graphs, diagrams, etc. Do not over-do pictorial representation of data just to get some colour into your dissertation. A small table for example, may better show the age profile of people, than a brightly coloured pie chart using half a page of space. Figures and tables should be numbered, and prefixed by the number of the chapter in which they appear, e.g. Figure 2.3 will be the third figure in Chapter two. The title and content of figures or tables should be such that they can be understood on a stand-alone basis. The reader should not have to browse other sections of text to gain an understanding of a figure or table.

## 1.4 Possible subject areas for your dissertation

The topic area that you choose for your work should ideally be related to the specialism that you are studying within construction. You should consider all parts of the construction process, from and including inception (clients with ideas that require projects), through to construction, maintenance, refurbishment, demolition and recycling. Most disciplines are keen to use their skills to improve the service provided to clients at all stages of the process. In practice, modern methods of procurement integrate the supply chain, and therefore all professionals are now involved earlier and later in the process than has traditionally been the case. You may consider issues from the perspective of any party in the supply chain, e.g. clients, end users, consultants, contractors, sub-contract specialists, suppliers, manufacturers, or indeed other stakeholders such as investors or the public.

Topic areas often include 'soft' people issues, such as job satisfaction, grievances, employee turnover, or quality of life measures. Resources such as sub-contractors, plant, material and capital (money) are also popular. You may want to specialise in finance, planning, legal issues or contracts, procurement methods or use of information technology. In the context that you may wish to consider variables in your study, popular dependent variables align with Key Performance Indicators promoted by Constructing Excellence in the Built Environment, e.g. client satisfaction, cost predictability, time predictability, quality, safety, etc. Sustainability issues driven by the climate change agenda are often researched. There is great potential for studies in many areas related to sustainability; in the UK, Building Research Establishment Environmental Assessment Method (BREEAM) and Code for Sustainable Homes can be used to identify many potential topic areas in this field, for example renewable energy. Defining and measuring best practice in a given field, may be the basis of a useful study. The definition of best practice could be an objective of your study met by the literature review. You may find doing this useful to you personally, since it is a valuable way to enhance your own knowledge in the field. The measurement of compliance with best practice by organisations or individuals may then be the basis for another objective, to be met by the main data collection process in the middle part of your study. When Paul Morrell took up the newly created post of UK Government Chief Construction Advisor in November 2009, he stated 'we're going to need to start counting carbon as rigorously as we count money, and accepting that a building is not of value if the pound signs look okay, but the carbon count does not' (Richardson 2009) - lots of opportunities, therefore, to measure carbon. It is not the intention that you should produce a 'project' of a descriptive kind, such as a design project. The emphasis is on data collection and analysis, around objectives. It may be management, technology or science-based.

Most often, part-time students select a problem from their workplace. Alternatively, you may select something that is current in industry or academia. You should have been reading about current issues throughout your study; as you are selecting the topic area for your dissertation, you should speed this reading up. The lead sources for current issues are web sites and conferences of your professional bodies, other academic conferences such as ARCOM (Association of Researchers for Construction Management), the weekly construction press and construction academic journals. To ensure your study has academic credibility, if you start from a practical perspective, you will need to take it back to its theoretical roots. Alternatively, you may start with a theory, and take it forward to its practical application; for example, flagship theories in management, such as leadership, motivation, etc.

## 1.5 Qualitative or quantitative analysis?

The middle of the dissertation should include some analysis – taking one element of a problem, breaking it down, establishing causes and effects. Robust analysis involves the application of some kind of academic tool; some academic tools may be considered more robust than others. The way you go about collecting data for analysis, and the way you perform the analysis, is one facet of a dissertation that distinguishes it from more conventional assessments or coursework.

Two major analytical schools exist: qualitative and quantitative. Crudely speaking, qualitative methods involve analysing words and quantitative methods involve the analysis of numbers. Some people may be able to use both methods, but sometimes a person is a specialist in one or the other. In their approach to a problem, researchers may lean towards methods which they understand best. Historically, there has been much debate between the two 'camps', to the extent that they have been described as wars. More recently, there is recognition that whether to use a qualitative or quantitative approach must be driven by the nature of the problem and objective to hand. The objective must drive choice of method, not the other way round. Ideally, you therefore need to have at least an appreciation of each method to allow you to select the best method to meet your objective. Mindful of your limited time, it may be that your research is predominantly based on only one type of data analysis. It does not matter which, provided there is some type of analysis of some type of data. In business terms, objectives are often non-negotiable; business executives or politicians will set objectives. They will then select the people with the appropriate methodological expertise who can meet those objectives. In your dissertation, it is possible that whilst objectives should drive methods, in reality it can be the other way round, and if you are setting the objective, you can set it to suit your strengths.

The boundary between the qualitative and quantitative can be blurred in questionnaires, by taking qualitative responses to closed questions, and coding them with numbers that have real quantitative value. For example, questions about product or service satisfaction may give respondents four qualitative labels as possible answers: (a) very satisfied; (b) satisfied; (c) not satisfied; and (d) not at all satisfied. The analytical process may allocate the numbers 3, 2, 1, 0 to the responses. The answers given are clearly qualitative; they express the feelings of respondents using a carefully selected word, e.g. I am very satisfied with this product. The passion intended by the qualitative response is diluted by the quantitative number. There are instances where boundaries between qualitative and quantitative are not blurred. Unstructured interviews, whereby interviewees speak freely, are clearly qualitative, and it would be inappropriate to allocate numbers to the data. Quantitative data (numbers) is evident in statistical data published in all spheres of business; government publish quantitative economic data, and employment data; companies have their production figures and profits.

Qualitative analysis aims to gain insights and understand people's perceptions of the world; people labour to give answers. Beliefs, understandings, opinions and views of people are examined. They may 'pour their heart out to you' about a particular problem. Expression and tone may reveal as much as the words. The data you receive is 'rich' and gives valuable insight. The data will be unstructured in its raw form. Such rich data could never be obtained in a tick box electronic survey. If your study was to seek out causes of dissatisfaction amongst some parts of the workforce, this is probably best done with at least some element of qualitative work. Data needs to be filtered, sorted and manipulated if analytical techniques are to be applied. Qualitative analysis may be merely findings derived from reading other work or from reading transcripts of interviews conducted by others; alternatively, it may be from transcripts of interviews conducted personally. Analysis may involve simple content analysis, whereby there is counting of key words, and deriving data meaning from high frequency hits. A more rigorous approach may involve some kind of coding and comparison of findings in a tabular format or content analysis (word counting) of data. The qualitative analyst will labour meticulously over transcripts. There could be analogies with criminal investigations where prosecutors labour over police interview transcripts; what did the suspect really mean when he/she said '. . .'. One may remember a testimony given by former President Clinton – what does the word 'is' really mean? A manual approach may be used to analyse data, whereby there is coding, photocopying, and then sorting by 'cut and shuffle'. Alternatively, analysts may use specialist or standard word processing software.

Quantitative analysis at its simplest level may be a mere comparison of figures. It may involve some descriptive statistics, such as calculation of means. It may include some medians, modes, or standard deviations. More rigorous statistical analytical methods involve inferential statistics. Most students should be able to understand and execute the simpler inferential tests; to do the tests is part of hypothesis testing. It requires that you understand the concept of variables; tests seek to determine causes and effects, and whether an independent variable (IV) influences a dependent variable (DV).

Some people may have an inherent fear of quantitative analysis or statistics. Be mindful that the use of statistical data is part of everyday life; at its simplest level, dealing with money. If you have a non-negotiable position that you will not do statistics, this is fine. But, if you do go down the qualitative route, be sure that you use robust data collection and analytical techniques as far as is possible.

Constructing Excellence may be used as an example of where qualitative and quantitative data merge or melt into each other. An array of qualitative data is collected to measure client satisfaction. The construction industry needs to know whether its clients are satisfied, and needs a measurement system to help it monitor its own performance. An output is needed that summarises data, gives quick comparisons, and allows executives to be able to quickly pick upon areas that need corrective action. The chosen way to do this is to use numbers. Client satisfaction, which has its origins in the collection of qualitative data, is therefore scored on a quantitative scale of 0–10. It is important not to allocate a number to client satisfaction 'anecdotally'. A robust set of criteria should be written that would allow a qualitative narrative to be scored within a specified range. The criteria may be in the form of expectations; what would be qualitative expectations of a score in the ranges 0–2, 3–5, 6–7, 8–9, or 10?

It is argued that the best studies comprise the analysis of both qualitative and quantitative data. The qualitative analysis may come first; speaking to people, teasing out issues and problems. The quantitative analysis follows, using numerical data to test hypotheses. The researcher may then revert back to more qualitative data gathering to help in interpreting results and findings from the quantitative tests. The review of the theory and the literature during the early part of the study may be considered to be a qualitative analytical tool, although the review may also include some quantitative analysis. Using both qualitative and quantitative approaches in a study can be called 'mixed methods' or 'methodological triangulation' (Clarke and Cresswell 2008). In the latter case, the three-sided analogy comes from two methods focused towards meeting one objective.

Miles and Huberman (1994) cite two strap lines at opposite ends of the qualitative/quantitative spectrum from Campbell and Kerlinger. Those who believe in quantitative work may have their view balanced by Donald Campbell, who stated, 'all research ultimately has a qualitative grounding'. Those who believe in qualitative work may have been influenced by Fred Kerlinger, 'there's no such thing as qualitative data. Everything is either 1 or 0'.

The key thrust of your analytical framework could be quantitative. There are possibilities to flavour your work and to create links into the qualitative school in two ways. The first is the literature review, which by default, is a qualitative appraisal, even though it may include some numerical data. The second is the possibility for you to speak to people, even informally (though better formally in interviews), at various stages of the study. At early stages it could be around helping you to define the problem and objectives; at later stages to interpret your results and develop conclusions. You can write up some brief notes that summarise your discussions and place them in an appendix. It would seem very inappropriate for you to jump in to a quantitative study, having based it on what you have read; desk bound studies can be insular and misguided.

It would seem reasonable that your work could be substantially, almost exclusively, qualitative throughout. However, this should not be to the complete exclusion of quantitative data, as though you have a phobia with numbers. You may draw on some quantitative data in the description of the problem or in the literature review, merely by citing some numerical data to perhaps justify the reason for your study. Further qualitative work does not preclude you from the need to have an appreciation of quantitative methods. Your approach may be to undertake an in-depth qualitative study that investigates a problem, teasing out potential variables. The closing part of the study is a recommendation, which sets up a study for another student, perhaps next year. The recommendation should be fully developed, to include definition of variables and the suggested quantitative analytical tools.

It is not a reasonable expectation that your dissertation should be based on the collection and analysis of substantive qualitative and quantitative data. You can only do what you are able to do, in the given time frame. Research programmes that are well funded may start with an in-depth qualitative investigation, followed by a hypothesis to be tested quantitatively. The whole research programme will close with some more in-depth qualitative work to help in the interpretation of quantitative results, and help in the development of conclusions. It may include a series of pilot studies at each stage, and some intervening qualitative work between the pilot and the main study.

## 1.6 The student/supervisor relationship and time management

There are lots of variables at large that influence whether dissertations will be successful or not. One of the lead variables may be the quality of the relationship you have with your supervisor. The student should be in the driving seat as far as is possible. It is just a fact of life that supervisors take different views of their role. Some supervisors point blank refuse to read interim draft submissions; that may be departmental policy. Some supervisors actively chase students. How much contact there is in a student/supervisor relationship seems to be important. At one end of the spectrum there are regular meetings, students regularly provide drafts of chapters and supervisors promptly read and provide feedback. At the other end, supervisors rarely or never see students other than for one preliminary meeting. A completed document then suddenly appears on the submission date.

So what level of supervision is appropriate? There is no definitive answer. The volume and type of feedback will vary between supervisors. Most will not identify mistakes in spellings or grammar; feedback may be of a strategic nature limited to such things as how the overall structure of documents can be improved, correcting fundamental misunderstandings in research methodology or in the subject area, and identifying gaps. It is absolutely clear that supervisors will not 'write' the dissertation for you. Key learning outcomes of dissertations are that students must demonstrate independent working and use their own initiative in developing their work. It is probably the case that most supervisors will not chase you; they are busy people.

Try to select your supervisor as soon as possible in the process. There is a limit to the number of students that supervisors can work with each year. Good supervisors may get fully booked at an early stage. Talk to more than one supervisor to try and tease out who has the expertise and interest in your topic area. Rather than a blank piece of paper and limited ideas, try to take a page or so to any meeting that outlines the problem you may wish to investigate. Demonstrate to supervisors that you have given some initial structured thought to your proposed study.

There are two key elements in the selection of your supervisor. Firstly, supervisors should have expertise in the area of your study. Secondly, you should have some personal chemistry with your supervisor. You may wish to ask your supervisor what the ground rules are in your relationship, e.g. frequency of contact, potential for feedback, expectation for interim submissions, preferred method of communication, e.g. e-mail.

A possible template that may be agreeable to supervisors is:

- You write a few pages to describe initial ideas and describe what the problem is to be investigated

- You meet with several potential supervisors and select the appropriate person
- You write regularly throughout the process. You submit drafts of your writing; your supervisor provides feedback—you redraft.
- You book face-to-face appointments with your supervisor; possibly every 3, 4, 5 or 6 weeks? You promptly write up notes of these meetings as an aide-memoire for the next meeting.
- You communicate informally with your supervisor, 'regularly'.
- You submit an almost fully complete draft of the dissertation one month before the final submission date; your supervisor gives feedback.
- You edit and proofread before final submission.

Note the repeating word 'you'. Whilst some of the above may be 'good practice' and part of the service that universities wish to provide, it may just not be the way you and/or your supervisor want to do it.

Whether or not you make steady progress with your work may be another factor/variable that influences whether your dissertation will be successful or not. Dissertations left to the 'last minute' are less likely to be of good quality. Last minute dissertations are often a tortuous journey for students; that is, they are constantly at the back of the mind as a worry and cause of stress. Dissertations which make steady progress are most likely to be better quality and an enjoyable journey. If you are not able to make good progress because of problems elsewhere in your business or personal life, you should keep your supervisor and personal tutor updated. Avoid getting to a late stage in the process before advising of your difficulties. University systems are generally paternal and supportive in cases of genuine difficulty outside your control. As part of your first proposal to your supervisor, you should prepare a bar chart or programme; this can be difficult, since to some extent, you are going into territory that for you is uncharted. There is no necessity to produce the bar chart from a critical path network, but you can loosely indicate with dotted lines activities which have some spare time or float. The following are some generic activity titles; try and adapt these titles so that they become bespoke to the context of your own work:

- Writing (include within the body of the programme completion dates for each chapter)
- Supervisor meetings
- Initial description of the problem and establish provisional objectives
- Networking to redefine the problem
- Final description of the problem
- Conduct and write-up the review of theory and literature
- Networking to establish definitive objectives
- Methodology design to meet the objectives
- Pilot study including data collection and analysis
- Data collection process
- Data analysis
- Networking to share findings, and develop discussion and conclusions
- Write-up discussion and conclusions
- Proofreading and correction
- Draft submission
- Final submission

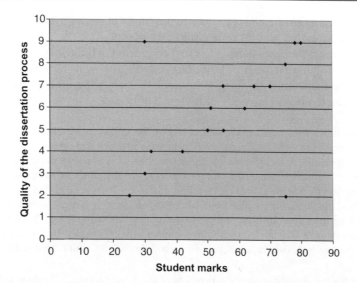

**Figure 1.1** Scatter graph of scores; quality of the dissertation process and student marks.

Figure 1.1 illustrates a possible outcome for a hypothetical research project; the objective is 'to determine whether the quality of the dissertation process influences student marks'. There could be several concepts wrapped up into ' . . . the quality of the dissertation process . . .'. There is just one measure of 'student marks'; that is the raw number on the scale of zero to 100. If the study were 'real', careful consideration would need to be given to defining each of the variables and to designing a method to measure them. The terminology used later in the text is that the independent variable (IV) is ' . . . the quality of the dissertation process . . .'. and the dependent variable (DV) is 'student marks'.

The results indicate that ' . . . the quality of the dissertation process does influence student marks'. Marks in the lower left cluster are poor process and poor marks. Marks in the upper right cluster are good process and good marks. There is an outlier at the top left; this is a good process but a poor mark—hopefully in reality this would never happen. There is an outlier at the bottom right; this is a poor process and a good mark. Perhaps this was an extremely talented student, able to work without supervision.

## 1.7 Ethical compliance

You should ensure that in your study you do not do physical or emotional harm to any person, including yourself. Obviously you will not do this deliberately, but you should not do it accidentally, thoughtlessly or carelessly. Construction is well practised at risk assessments to avoid physical harm. If your research is laboratory based or surveying fieldwork, you should undertake a risk assessment and include it in an appendix. This assessment will follow your university procedures, including best practice promoted by the Health and Safety Executive (n.d.)(working at height regulations), e.g. avoid, prevent, mitigate; and (a) look for the hazards, (b) decide who might be harmed and how, (c) evaluate the risks and decide whether the existing precautions are adequate or whether more should be done, (d) record your findings where necessary, and (e) review your assessment.

The possibility of emotional harm is less well considered in built environment disciplines. Your university will have its own code of practice or similar on its website, detailing ethical standards to be maintained in doing your research. It may be underpinned by two key procedures (adapted from University of Bolton 2006):

For all dissertations, irrespective of the subject area or data collection method, it is likely that you will have to complete a university 'permission to do research form', and agree and sign it with your supervisor. Appendix A includes an exemplar checklist of items that might be included in a 'permission . . . form'. If some items on the checklist identify issues or risks, you will have to describe how you plan to avoid, prevent or mitigate. The form may require that you indicate that you have read the university's ethical code, and that you agree to comply with it. Include your completed form in an appendix to your dissertation, and in your methodology chapter describe the process that you went through to ensure you complied with ethics rules. If there are some issues which are thought high risk, approval may be needed through an ethics committee.

At the point that you come face-to face with others, or ask to use data sets that belong to others, you should provide an 'information sheet' and gain written or verbal 'consent'. The information sheet is required so that there is absolute clarity and a record about the 'rules' of engagement. It should set out: (a) the purpose of the investigation, (b) the procedures; the risks (including psychological distress), (c) the benefits, or absence of them, to the individual or to others in the future or to society, (d) a statement that individuals may decline to participate and also will be free to withdraw at any time without giving a reason, (e) an invitation to ask questions, (f) contact details of your university department's Research Ethics Officer so that participants may report any procedures that seem to violate their welfare. Participants should be given plenty of time to study the information sheet, and consult relevant parties. You need to be respectful of other people's time, and it may therefore be appropriate to specify a time limit or target for your contact with the respondent. Keep in mind this is 'only' undergraduate study. Whilst university codes may require you to articulate the benefits of your research to society, it is probably sufficient that you merely state the benefit is to further your own education. In some cases 'consent' may be required by respondent signature on a pro forma. Alternatively, you might think it adequate to secure written permission perhaps by e-mail, or perhaps just verbal approval with a note in your diary or research papers.

You should not undertake questionnaires, interview or observe people in such a way that it might put respondents under pressure, instil anxiety or induce psychological harm. Neither should you do anything that people might consider offensive, even if only mildly offensive, e.g. asking about things that people may consider private, such as family details or personal or company finances. Issues around gender, culture, and religion should be treated sensitively. Obviously you will not convey untruths to people in your research, but making judgements about the issue of deceit can be more difficult. For example, is it appropriate that you observe people in their place of work without advising them you are collecting data for a research project? You will probably be forced to do this if you are 'external' in a given situation, but careful judgement needs to be made if your data collection is made alongside your normal work activity, e.g. a study about the effectiveness of site meetings in which you are routinely involved as a participant as well as in your role in the same meetings as a researcher. In this case, should you ask permission of participants to tape record meetings or take notes, or should you just ask permission of the Chair, or should you just do it without anyone knowing?

Since this is not a public meeting, especially if there is to be a tape recording, it is probably best that the consent of the meeting is sought through the Chair. Also envisage a situation where you may wish to record observations of craftspeople whilst they work. You may do this as part of your routine employment, or you may spend extra time observing with their knowledge, or perhaps observe from a vantage point without their knowledge. 'Deceiving' people is sometimes justified, providing that the value of the research outweighs the principle of gaining permission. You would have to substantiate this in your 'permission to do research form'. Your argument may be that if respondents know you are collecting data for research, it would unduly impinge upon study validity, and prevent tangible benefits being realised.

You must be particularly careful if respondents are considered vulnerable. Construction designers may legitimately wish to communicate with vulnerable users of buildings and other parts of our infrastructure. If your work involves children or young adults, perhaps talking to 14 or 15 year olds about their perceptions of construction as a potential career option, you should take on board your supervisor's advice. People who are in training, or who are accountable to you at work such as craftspeople, may feel their job positions are threatened by your wish to collect data. People more senior than you may be fearful about data reaching their superiors or the public domain. People with a disability or learning difficulties may be reluctant to talk about adjustments they need for their work. People in organisations that are downsizing may be reluctant to talk about anything, and if that is the case, you should respect their wishes.

Part-time students may be able to get data from their own organisations. If it involves other people, permission should be sought from their employer; if it involves taking data from current or archived files, again permission should be sought. If your research involves collecting data from patients, carers or staff in the National Health Service or social services, e.g. about the merits of various building designs in patient recovery, you should gain approval from those agency research ethics committees. Confidentiality of data should be maintained; that is, data should not be used for any purpose other than that stated. Anonymity of individuals and companies should be protected, by typically referring to person A or organisation B. If you are using a data set about a company that is already in the public domain, e.g. the profit figure for a public liability company, it is acceptable to name names, providing the name cannot be used to find a link to other confidential information. You should offer participants in your study access to a summary of your findings; often individuals do not take up that option, but if you are using data from your workplace, your employer may be interested.

There are examples where you do not need to give out information sheets or gain consent. In the event that you use a postal or electronic survey, consent may be implied by its completion and return, thus removing the need for written consent. If you wanted to observe in a public setting how people use a particular facility, that is ethically acceptable, but if you wanted to talk to them, you should provide an information sheet. If there are any adverse events during your research, involving either yourself or others, you should report these immediately to your supervisor. You should also be mindful to comply with the Data Protection Act by, for example, storing any data set that may identify individuals securely, and disposing of it appropriately (shredding not bins).

Ethical rules should not be viewed as a deterrent to you being proactive in your research, and then alternatively encouraging you to undertake passive desk studies. If you follow the prompt list style of the check lists in permission to do research forms, that should prevent you from making a mistake. Your research permission form is good practice, morally appropriate and is your plan; your information sheet helps your respondents and hopefully puts them at ease. If you are able to grasp the principles of research ethics, there is the potential for lots of spin-offs into other spheres of your personal and professional life.

## 1.8 House style or style guide

Your dissertation needs to be accessible to all potential readers, and its style needs to be consistent within itself. Some readers have disabilities that make conventionally presented text inaccessible. Some disabilities affect a relatively small percentage of people, e.g. partial or whole sight loss. However, dyslexia is relatively common, affecting nearly 10% of students in higher education. You need to be mindful of how you set out text, font style and font size.

For many people with poor sight, well-designed printed documents using a minimum of 12 point text is enough, although the Royal National Institute of Blind People recommends 14 point, to reach more people with sight problems. People with dyslexia usually find sans serif fonts such as Arial easiest to read. Left justification is thought to be better than full justification, since the latter necessitates that the spacing of letters within words open to take up the full width of lines. This exacerbates a feeling that some people with dyslexia have, of words merging into one another.

It may be prudent to place a CD-ROM version of the dissertation in a plastic wallet stuck to the inside cover. An electronic version allows people to increase font size or use other sophisticated software to read documents. In the event that your CD is not used, you would at least be demonstrating to examiners that you have empathy with difficulties experienced by others.

Your university may have its house style that you must follow. Hopefully, it will not be too prescriptive, thus leaving you with a degree of flexibility and personal choice. You should aim for absolute precision in use of the style. Even small differences within the same document can frustrate readers, and distract their concentration from their lead task of digesting subject material. Adapting to the discipline of house styles is good practice for industry. Most businesses have their styles, so that they have consistency in the way they present information to clients and customers. If you are a part-time student, you may find it frustrating to have to learn two house styles; one in your workplace and one at university. Since the real issues are about accessibility and consistency in documents, and not about the style itself, your university may be happy to let you use your workplace style. If you want to emphasise to examiners that you have not put your document together without regard to accessibility and consistency, write a precept in the preliminary pages stating you have followed a style, and then put details in an appendix.

If you are not required to follow a style, and do not have one in your workplace, one is suggested below, based on disability literature (ABECAS 2005):
Use the following:

- Font type: Arial or Times New Roman
- Chapter headings: 14 pt bold, sentence case
- Subheadings: 14 pt bold, sentence case
- Main text: 12 pt line and a half spacing, sentence case

- Table text: 12 pt single line spacing, sentence case
- Margins 3 cm all round
- One and a half line space between paragraphs, not indented
- One and a half line space between headings and text, text and new heading, and tables/figures and text
- Page number at centre bottom of each page, 12 pt
- Left justification with ragged right edge
- Allow generous spacing generally within documents
- Black or dark blue print colour

Generally, avoid the following:

- Centred text except for headings
- Upper case fonts (or capitalisation)
- Italic fonts (bold is a better form of highlighting)
- Underlining
- Roman numerals, e.g. (iii), (iv), (vi), and numbering 3 and 8, 6 and 9, which are difficult to recognise. Bullet points or (a), (b), (c) are preferred

## 1.9 Writing style

You are likely to have a word limit or word guide for your dissertation; perhaps 10 000 words. Whilst initially this may seem to be an enormous amount, it is not. A draft of your document may be say 20% higher than your final document, and you will have to work hard to remove irrelevant information. A rule to follow is, 'if what you have written should be in, it should be in; if it should be out, it should be out'. Following this rule requires a robust editing process. It follows that you must write concisely and with precision. Every sentence must read with absolute clarity – it should mean to the reader exactly what it means to you as the writer.

This is not a textbook in writing style or grammar, but here are some tips arising from common mistakes by construction students:

(1)  Do not write in the first or second person. The following first person words should not appear: I, me, mine, my (singular), our, ours, us, we (plural). Nor the following second person: you, yours, your. Whilst they are all acceptable in informal writing, they are not thought to be professional writing styles; a passive approach is required. If the first person slips into the first draft of writing, it may be easy just to rearrange sentences. For example a first draft reading 'we need to improve the safety performance of the construction industry', can be rephrased 'there is a need to improve safety performance of the construction industry . . .'. On the occasion that you may need to refer to yourself, you can be author or the writer, not 'I', e.g. 'this study arose from the author's experience working in private practice . . .'

(2)  Avoid writing in the singular; do write consistently in the plural. Therefore, instead of a manager, a contract, a client, a sub-contractor, etc., write managers, contracts, clients, sub-contractors, etc. This can be really difficult, even with practice, but it does have advantages, thus:

(a)     It helps keep writing concise. A sentence that reads in the singular 'the project manager is responsible for the quality on the site', reads more fluently and with four fewer words in the plural as 'project managers are responsible for quality on sites'.

(b)     It can eliminate inappropriate mixings of singulars and plurals, so that a sentence written 'an architect should be the chair of a meeting and they should ensure ...' (a singular architect is not a they, but a 'he' or a 'she') is better as 'architects should chair meetings and they should ensure ...'.

(c)     It almost fully eliminates the need to address the gender issue. Some people are quite happy to call a manager 'he', whilst others may be offended. You do not want to offend anyone, therefore avoid the use of 'he'. In the singular, you may write 'the architect should chair the meeting and he should ensure ...'. In the plural this becomes 'architects chair meetings and they should ensure ...'. You can write in the singular 'an architect should chair the meeting and he/she should ensure ...' but this attracts readers' eyes to the gender issue. It is clearly acceptable to use the male or female gender if you are talking about an individual, e.g. 'Mr Smith chaired the meeting and he ...'. In your literature review you may cite Smith (2010), and 'he stated ...'; but be sure it is 'he'. It is best not to get too hung up about the gender issue. Whether you write, for example, craftsmen or craftspeople, it is not likely to make any difference to your mark. UK statute routinely uses the 'he' to also mean 'she', using a precept such as 'any reference to the masculine gender shall be taken to include the feminine'. If you are most comfortable writing of craftsmen, you can do so. Deflect the potential for criticism by including a precept in your preliminary pages or introduction chapter. But otherwise, still try to avoid gender in your writing where possible.

(d)     It promotes consistency in your writing. It some cases it will be inevitable that you write in the plural. It would be a contradiction to write in one paragraph of a singular surveyor, and then in the next of plural projects.

(e)     Finally, the purpose of research is often to seek out how the population at large behaves. It would seem a contradiction therefore to write of the population as though it comprised a singular company, person or project.

(3)     Inappropriate mixings of singulars and plurals also occurs when students write of organisations like 'a government, 'a professional body', 'a company', etc. using the plural 'they'. Each is represented by many singular people coming together as 'they'. Whilst the latter is correct, individually, these organisations have a singular legal identity. A sentence that reads with the plural they, 'government decided they would change the law ...', may better read with the singular it, 'government decided it would change the law ...'.

(4)     Avoid colloquialisms (not formal or literary), slang and nicknames. Some of these may arise from the construction industry itself or local dialect. Not all readers may understand them, e.g. a mobile elevated working platform (MEWP) is informally called a 'cherry picker'. Not everyone may know that, and you should therefore formally call it a MEWP. If the context of your narrative is that you need to write of a 'cherry picker' or use slang or nicknames, that is fine, but put them in speech marks to illustrate this is not part of your professional writing.

(5)   Avoid using words abbreviated by apostrophes, e.g. can't, won't, don't. Write them out in full as cannot, will not, do not. These sorts of abbreviations are acceptable if you are taking them verbatim from other sources. If the latter is the case, again include them in speech marks.

(6)   Do not use any bad language, mild or strong. In qualitative work it can be important to understand the passion or anger that people may feel about some issues. People may use bad language to express their feelings. Passion or anger must not be lost in analysis. You may type up verbatim transcripts of interviews, and include them in appendices. Some readers may be relaxed about bad language that is left in. Other people may be offended, and you leave yourself open to criticism. You do not wish to offend anyone. Therefore, whether in the main body of your dissertation or the appendices, take bad language out. Substitute it with an appropriate number of dashes. Readers will understand.

(7)   Make sure you can use apostrophes correctly. The two key uses are to indicate possession and to indicate missing letters in the middle of words. It is the possession issue that causes students most problems. For example, 'the house of the architect'; by deleting the word 'of' (the possessor) and switching house and architect around, it becomes 'the architect's house'. The apostrophe is used in place of the word 'of'. If the house is owned by more than two architects, the apostrophe is after the s, thus 'the architects' house'. If a house is owned by an individual or family with the surname ending in 's', the latter 's' can missed off so that it is 'Jones' house' not 'Jones's house'. If you are not sure how to use apostrophes, find out. There are lots of university guides and useful websites that you can locate through search engines. Imagine the discussion that may take place between two examiners who disagree about the mark for a dissertation; one argues for a high mark, whilst the other suggests a lower mark stating 'this student cannot even use apostrophes correctly'. Imagine employers discarding CVs because candidates cannot even use apostrophes correctly.

(8)   Make consistent use of numbers written as figures or as words, that is the word 'one' or the figure '1'. By convention, anything less than 100 is written as words. You may be uncomfortable writing ninety-nine, and prefer 99. In that case, write numbers up to ten as words and over ten as figures. Be sure to be consistent, and not to write '12' in one sentence and 'thirteen' in the next. Numbers used in calculations should of course be kept as figures; similarly ages, measurements and percentages - 10% is better than 10 percent.

(9)   Which writing tense to use often causes difficulty. There are four types of tense: (a) past tense, which is sub-divided into imperfect, perfect and pluperfect; (b) present; (c) future; and (d) conditional. In simple terms people often refer to only three: past, present, and future. Students often write without regard to tense and inter-change tenses within adjoining passages of text. You need to be mindful of the tense in which you are writing at each stage of the document. The abstract is written as a concise summary after the whole of the document has been completed. What tense is appropriate?

- Present tense 'the objective is ... the method is ... it is found that ... it is concluded that ...'?
- Past tense 'the objective was ... the method was ... it was found that ... it was concluded that ...'?

- Or a mixture to show that the beginning of the study is past, but the end of the study is present, thus 'the objective was … the method was … it is found that … it is concluded that …'?

The introduction chapter will be written as the document proceeds. Some parts will be written at the beginning of the process, and some towards the end. The introduction is setting out what will appear in future chapters, but it is also telling the story of what has happened, in a real time frame, in the past. What tense is appropriate - 'the objective is … the method was … Chapter 2 will outline …'?

The literature review is about work in the past. If writing about work that is well established and dated, the past tense will usually be used, e.g. 'Taylor (1911) stated that …'. However, if the work is more recent, it may be written as 'Smith (2009) finds that …' or 'Smith (2009) found that …'. On the one hand, the methodology chapter tells the story of what you have done in the past, but it also describes what will be presented in the middle of the dissertation. Therefore, it may be written as 'a survey of existing buildings was (past) undertaken, and the data recorded will be (future) analysed in the next chapter …'. The analysis chapter may use the present tense; the result is 'this' and 'this is' found. Discussion about the results and findings may be in the past tense – 'in the last chapter the result was … and it was found that …'. Conclusions may start to arise in the discussion chapter, and may be in the future tense 'it will be concluded that …'. As you get to the end of the document you may revert to the present tense 'it is concluded that …'. There is no definitive answer about which tense should be used. The important point is that you should pick your tenses deliberately, recognising in your own mind instances where you have made choices, and remaining consistent in those choices.

## 1.10  Proofreading

Proofreading is an essential part of the dissertation process. It must be thorough and meticulous. It must cover content, grammar, spelling, apostrophes, layout, and presentation. You should aim for perfection. You should be confident about putting your dissertation on the desk of chief executives or other captains of industry. You should proofread your own work as much as possible. On the one hand you may aim to get chapters 'signed-off' as though floors on a building that are being handed over. But, when writing subsequent chapters, you may need to go back to signed-off work to make amendments. Part of the process is that you will be re-reading and re-reading your work again on a continuous basis. You must continue to polish it and polish it again. A time gap between reading helps. If you 'finish' a chapter and then go back to it a week or so later, you will no doubt find many things you want to change to make it read with greater clarity or text to delete or add.

If you are doing this thoroughly, there will come a time when you are just too close to your work. There will be mistakes that you will never find. To get the perfection required will need the independent help of a proofreader. Proofreading is a profession in its own right. It is not expected that you will seek professional help. The proofreader may be a member of your family, friend, or work colleague. Construction expertise is not necessary, since hopefully you will have been networking with construction professionals and your supervisor throughout the process. There is a line that you must not cross with proofreading by others. Proofreading is proofreading; it is not for someone else to rewrite your work. Students with dyslexia may be able to substantiate a case for greater support in the proofreading process through their university disability advisors. Whatever support you do get with proofreading, declare it in your acknowledgements in the preliminary pages. In the final stages you are building up to the final print off. In reality there may be several 'final' print offs. The final print off may go to the proofreader, who may find lots of typographical errors. Be mindful that relatively minor changes can lead to substantive disruption to document presentation. In your final, final document, graphs, pie charts, histograms, photographs, or maps should be colour printed.

## Summary of this chapter

The dissertation is the flagship document in your degree. Language used to describe research goals can be fuzzy. However, clarity in your objectives is essential. The opening part of your dissertation may be an introduction and literature review. The middle part may describe your methodology and present your analysis/results. The closing part may be discussion and conclusions. When selecting a topic area, you should speed up your general reading around current issues. You may choose to do qualitative or quantitative analysis or a mixture of both; whichever is used, the analytical tools must be robust. You should be clear about the way you and your supervisor will work together. The presentation style of your document must be consistent within itself, and you should consider readers who may have partial sight, dyslexia or similar. You must read and understand your university ethical codes; do not harm or offend any person during your work. You should try to complete your dissertation one month before the final submission date, and use the last weeks to make improvements that may be suggested by your supervisor and for proofreading.

# 2 The introduction chapter to the dissertation

The objectives of each section of this chapter are:

2.1 Introduction contents: identify the parts required in an introduction
2.2 Articulation or description of the problem and provisional objectives: to emphasise that a well-developed narrative around a problem provides the best foundation for a study and helps to identify provisional objectives

## 2.1 Introduction contents

Writers need to focus very carefully on what an introduction should contain. Most academic coursework may contain an introduction. The introduction is not a bit of a chat or the first part of the whole. *The Oxford Dictionary* defines an introduction as 'an explanation section at the beginning of a book'.

For a dissertation, the introduction should tell readers where the rest of the dissertation or the chapter is going—what items will be covered. Having read the introduction, readers should have a sound platform from which to continue reading. There should be no surprises in remaining chapters, and it is just for readers to continue their read to gain greater understanding and ascertain details. You should start writing your first draft of the introduction chapter at an early stage, but you will not be able to complete it until near the end, since it reports on what you have done in the latter stages. Students often write short introductions; this should not be the case. They should be a substantive part of your document. The introduction chapter should include the following:

(a) Introduction to the introduction.
(b) Definitions of important phrases; this may expand on the definition of one or two key phrases which are fundamental to your work. The definitions should again be drawn from authoritative sources, and may bring together conflicting definitions. For example, a dissertation that examines 'sustainable development in construction' may start to

*Writing a Built Environment Dissertation: Practical Guidance and Examples.* Peter Farrell.
©2011 Peter Farrell. Published 2011 by Blackwell Publishing Ltd.

define this as: 'development that meets the needs of the present without compromising the ability of future generations to meet their own needs' (Brundtland 1987).

(c) Background to the topic area and definition/articulation or description of the problem.

(d) Possibly a historical perspective: this need not be an in-depth narrative, but it can be useful to readers to establish a context for the study. Subject areas are rarely 'new', and they may have origins going back many decades in UK construction, or internationally or in other industries. For example, many initiatives in construction have their roots in manufacturing.

(e) Research goals: the aim of the study, research questions, objectives and hypotheses. 'Research goals' is a useful umbrella label for all these terms.

(f) An outline methodology: it will give a brief summary of the methodology and may include a brief diagram or model of the variables being measured. The model should link the aims, objectives and hypotheses. The methodology as described in the abstract may be two or three sentences. The outline methodology as described in the introduction may be two or three paragraphs. There should be a full chapter devoted to a detailed description of the methodology later in the document.

(g) An outline of the remaining parts of the dissertation: this will tell the reader how the rest of the dissertation will develop. It is necessary to explain the intended structure and the route that the work will take. It will briefly tell the reader about the contents of each chapter.

(h) A summary.

## 2.2 Articulation or description of the problem and provisional objectives

There is agreement that research in construction is often problem-based; this is supported by Creswell (1997, p.74) and Silverman (2001, p.5) in the fields of human and social sciences. However, some terminology in texts is variable, for example according to Hart (2001, p.9) the first part of a dissertation should be a rationale, Holt (1998, p.9) asks for a broad discussion, Naoum (2006, p.14) suggests a purpose, and Fellows and Liu (2008, p.42) a proposal. This does not matter, since a definition of all these terms usually includes 'articulation of the problem'.

One of the key skills of higher level study is problem-solving; indeed it may be argued that it is 'the' primary skill. The examination of an industrially-orientated problem helps to integrate academic and industrial communities; in many cases sponsorship of academic work is only offered if output can be fed back into the market environment. Articulation of the problem is the first starting point for a research project. It provides the foundation for the study. A really well thought out narrative at this stage gives an excellent platform for good work. It should be the subject of many iterations, so that improvements are made as the work progresses. It may be based on a modest amount of reading, and will be subsequently redefined as the researcher becomes more knowledgeable. Part-time students may wish to draw on problems from their own work environment. Collecting data for analysis at a later stage of the study can be difficult in research; it follows that if the problem is based in the workplace, you may have relatively easy access to data. Any evidence found to support the articulation of the problem should be cited. Redefinition may result from some exploratory interviews or from the early stages of the literature review. A key concept in research is to 'circulate'; that is, get out and talk to people about your study. Do this informally. If you talk to students or academics, ask them about their research so that you can learn about broader concepts and methods of research. Talk to people at all stages of the study. Examiners may wish to see that students have been pro-active in their research, rather than passive. Do not be shy. Desk-bound studies can be labelled dangerous, insular, blinkered and not valid. There are many forums in which to talk informally with knowledgeable people in your profession. For example, professional body seminars often welcome students, even if they are not members. So—leave your desk, be proactive and travel.

Take the first drafts of your narrative to your supervisor and other people for feedback. Undertake some formal unstructured interviews; maybe two. The basis of these interviews may be 'look, here is the problem as I see it, what do you think?' Ask about possible omissions in your work; have you fundamentally misunderstood something? You may get some leads to important publications or other authoritative sources that you would otherwise miss. Re-draft your articulation of the problem after your interviews. When you write up your research methodology, describe how you did some initial reading and writing, and how then you undertook some interviews, before re-drafting your work. This will all help in supporting the 'validity' of your work.

It is unlikely that students will have the resources to take on both robust quantitative and qualitative analytic methods within their work; that is, the triangulation or mixed-methods approach, as suggested by Somekh and Lewin (2004, p.274). Whether the main analytical method is quantitative or qualitative, the validity of the work can be improved and flavoured by speaking to people.

It may be useful to put the problem in the middle of a continuum. Before the problem or to the left are the perceived causes, the things that may have an influence on the problem. In the middle is the problem itself, and the output or to the right of the continuum are the consequences of the problem - what is it that results, or how does the problem manifest itself, what are the effects? There is some reflection and early speculation about cause and effect. Using safety as an analogy to illustrate this, possible causes of safety problems are lack of investment, training, education and poor safety culture. Safety problems manifest themselves in incidents and accidents. It is useful to support the problem with some statistics, e.g. considering the national or international picture, the number of accidents, the number of lost working hours. The consequences are losses of time/money in the production process, negative image labelling for industry, and of course not least, the human consequences for victims and their families. The research process often involves finding evidence to prove that perceived causes of problems are actual causes; it is to seek out the variables. In business terms, if there is evidence that a known cause has a known effect, that known cause can be 'managed' or 'manipulated'. The proof or evidence found should then be fed into the research conclusions and recommendations. The thought process at this time may be around the concepts of cause and effect, but as the study progresses, the language of cause and effect may change to variables; that is, independent variables (IV) and dependent variables (DV).

Initial articulation of the problem need not be in too much depth; perhaps two or three pages will suffice, but it can be added to as you enhance your knowledge at the literature review stage. It may be beneficial to conclude the articulation of the problem with one, two or three provisional objectives. These objectives are written in the knowledge that they are likely to change as work progresses, since your knowledge in the topic area is still being developed. They will, however, give you some early focus, and they are useful for recognising change in the study as it takes place. The objectives should be constantly under review until the end of the literature review, when they should then be 'written in stone'. The articulation of the problem, the final output being the result of many iterations, should also be 'written in stone' at the end of the literature review. The task for researchers beyond the literature review is to 'merely' meet the objectives that are set, employing an appropriate methodology.

The articulation of the problem should come before writing provisional objectives. Students often comment at an advanced stage of their studies 'I am not sure what my objectives are'. If the articulation of the problem is well thought out and developed, the objectives should 'hit the reader in the face'. They will be obvious. An alternative approach to writing the objective at this stage is to decide what you want to measure. Measurement is an important concept that underpins what you will do in your dissertation. As the study progresses, you may develop the idea to measure two things—an IV and a DV. Perhaps it is sufficient just to think about measuring one thing at the moment, e.g. compliance with best practice initiatives in health and safety in construction. That would seem a useful thing to measure. How does the UK construction industry score on a scale of 0–10? We would hope 9 or 10, but perhaps the answer is only 6 or 7? Do the research and find the answer to this question. A sample narrative of a problem is given in Appendix B.

## Summary of this chapter

The introduction chapter sets the scene and tells the reader how the remaining parts of the document will develop. It should be a substantive part of your dissertation. A well-defined problem that is based upon several iterations will provide a strong foundation for the study and help to identify objectives. You should take first drafts of the description of your problem and provisional objectives to others, and ask them what they think.

# 3 Review of theory and the literature

The objectives of each section of this chapter are:

3.1 Introduction: to provide a context for the literature review
3.2 Judgements or opinions: to distinguish between the authority of each
3.3 Sources of data: to distinguish between and to identify the most rigorous material
3.4 Methods of finding the literature: to identify the tools to be used when searching
3.5 Embedding theory in dissertations: to encourage students to address theory in their dissertations, and to distinguish between theory and literature
3.6 Referencing as evidence of reading: to emphasise the necessity of reading and citing
3.7 Citing literature sources in the narrative of your work: to illustrate how to 'cite whilst you write', and how to cite verbatim or by paraphrasing
3.8 Secondary citing: to give examples of how to distinguish in your writing between primary and secondary sources
3.9 Whom to cite in your narrative: to give examples of citing authors and organisations and to show how to emphasise the authority of sources
3.10 References or bibliography or both: to distinguish between and promote both
3.11 Common mistakes by students: to provide examples
3.12 Avoiding the charge of plagiarism: to explain what it is and identify software to help

## 3.1 Introduction

The literature review is an important part of dissertations. You need to survey previous work to determine whether similar work has already been executed, otherwise you cannot be assured your work is original. The literature review focuses almost entirely on the work of others. You need to be looking for: (a) similarities to your work—what can you draw on to help you, and what is it that is slightly different that you may do?; (b) gaps in the literature; and (c) contentious issues—do authoritative sources disagree? In many spheres of life this occurs frequently in both the hard and soft sciences. Sometimes disagreements will be voiced 'loudly' in the literature, where one source produces 'evidence' with the intention of contradicting

*Writing a Built Environment Dissertation: Practical Guidance and Examples.* Peter Farrell.
©2011 Peter Farrell. Published 2011 by Blackwell Publishing Ltd.

'evidence' produced by another, e.g. in the climate change debate. Sometimes the disagreements are not so apparent, and you may have to work hard to tease them out. Is there a facet of your topic area that has not been examined before at all?

There is rarely an argument to say literature reviews are not necessary. This could only be justified in circumstances where your work is closely associated with a previous research project. The literature review may have been completed in that previous project, and the thrust of your work may be to substantially develop one strand of what has already been done. In such a case it would be at least prudent to update the literature review of the earlier project.

A question often asked is 'what percentage of the dissertation should be devoted to the literature review?' The percentage may be based on time devoted or words written; students probably ask the question in the context of the latter. There is no definitive answer. It is normally a substantial piece of work. The suggested structure of a dissertation is six chapters. The literature review may be Chapter 2 of six; it may be part of the opening third of the document. Perhaps it should be more than one-sixth of the dissertation? Up to one-third of the document may not be unusual, but there should be one over-riding caveat: quality not quantity.

Often students select an area of study because they want to enhance their knowledge in a field. The literature review may be the part of the dissertation where you can do this. You may specify a study objective to be met by the literature review such as to review current theories, or to baseline current knowledge, or to establish current best practice. You hopefully become immersed in your subject area as you develop the thirst for knowledge. The literature review involves the compilation of a large number of articles and extensive reading. You may save some articles electronically; for others you may want to develop a well-organised paper file. Students often have lever arch files of previous papers where they have highlighted important text.

Holt (1998, pp.65–90) uses the extremely useful analogy of considering the review of theory and literature as though a funnel; a wide top part, moving down the aperture towards a narrow outlet. To develop his analogy, the wide top part is a receptor for material from a wide variety of sources. It should also consider the topic area in an international setting and across all professions or industries. As you move down through the funnel, material is considered in the context of the country of study, be that the UK or elsewhere, and in the context of the construction industry. Finally, at the narrow outlet there is material that will be related to your study objective. You may find yourself starting at the wide top, and then you need to work hard to find a narrow field of literature directed towards a modest objective. Or alternatively, you may start at the narrow outlet, and then you will need to work upwards to put the study in the context of broader fields of knowledge. It is important that you address both ends of the funnel. A study that remains at the top is likely to be too superficial to be of value; a study that is only at the bottom will fail to put itself into strategic context. Consider starting at the top of the funnel with a study to investigate climate change; as a topic area in itself it is too broad for a mere dissertation, but the international context must be addressed by reference to international authoritative publications and agreements. The position of the UK government can be identified and appraised against the position of a variety of pressure groups. There can be links into the Code for Sustainable Homes and Building Research Establishment Environmental Assessment Method (BREEAM) ratings; also consideration of how insulation and renewable energy sources can be used to reduce $CO_2$ emissions. The role of energy generation and transmission industries may be briefly examined. The narrow part of the review may be around the potential use of intelligent energy management systems by speculative developers or a comparison of energy generated by solar and wind or ... or ...

Readers will be able to deduce from the review what the latest up-to-date position is in this field; what is the extent of current knowledge, what is happening at the leading edge. It should be a summary of the state of the art. There can be some limited explanatory or descriptive material to provide context for readers, but not too much. Too often literature reviews are a mere description of what has gone before. There is an assumption that readers are construction professionals, but not necessarily experts in the topic area you have selected. The review should bring together common themes and issues in the literature, make intelligent links and demonstrate that the literature has been examined with insight. Most of all, the literature review should be critical and judgmental; it is these latter concepts that are often most difficult for students to master. It is not adequate to re-present the work of authoritative people, and then merely rubbish it. It is not sufficient that a literature review merely comprises statements extracted from previous work, which are bolted together in a clumsy fashion. The thrust should be to collect evidence from previous work, and pitch the work of one or several sources against others in a critical sort of way. Thus the criticism is informed. The critical nature of a literature review may be enhanced by the use of terms which emphasise comparison is taking place, viz.: whereas, on the other hand, alternatively, but, another view, the opposite stance, this is contradicted by. As you write, these sorts of terms may help to focus your attention on the need to be critical. But try not to use these phrases 'mechanically'. The literature review needs to flow as a 'narrative' and as an enjoyable read, and you need to demonstrate your writing skills as you weave it together. Take care not to throw in anecdotal statements, even if they are from your practical experience, as though they are facts, e.g. 'the construction industry is inherently unsafe'. As Fenn (1997) correctly states 'prove it'; where is the evidence for this bold assertion? You should not be discouraged and made afraid to write, since such statements may help you to weave the narrative together. You must either find the proof or if appropriate use a caveat thus, 'it may be argued that the construction industry is inherently unsafe'. Using another example, it may be tempting to write as fact $CO_2$ causes climate change, since everyone 'knows' this. However, this type of writing is not acceptable without citing the evidence so that the reader can go back to the source if necessary and challenge it. What is acceptable is 'carbon dioxide ($CO_2$) is one of the main greenhouse gases which causes climate change (Energy Saving Trust 2010)'. As part of your review you should also identify authoritative sources that dissent from this view.

Towards the end of the review you should assess, in your words, the implication of the literature on your study—relate it to the aims and objectives. It may be titled 'discussion', 'summary', 'critical appraisal' or 'appraisal'. Whilst the literature will be from a wide variety of sources, it must be written as though it were a funnel, with the output being consolidated and narrow. Legitimate contentions, assertions and arguments for advancing the area of knowledge further should be given. The identification of gaps is justification for further research, and should therefore lead to the objectives or hypotheses of the new research project. You should identify any fundamental issues that arose. Revisit the research questions, objectives and hypotheses at the end of your literature review. Having done your reading and appraisal, you have transformed yourself from having 'too little' knowledge to being an expert. Provisional objectives set at the start of your study should now be reappraised and adapted as necessary; at the completion of the literature review the objectives should become 'set in stone'. It is merely for the remaining parts of the study to meet these objectives.

A sample literature review is included in Appendix C.

## 3.2 Judgements or opinions?

*The Oxford Dictionary* defines an opinion as a 'belief or assessment based on grounds short of proof'. Judgement is defined as 'critical faculty, discernment, insight.' It may be helpful to consider the exercise as one whereby all the evidence pertaining to a particular issue is collated and placed on a set of balancing scales. Some of the evidence may be substantial, robust and stem from a research project which has been well resourced and funded. Other evidence may be less substantive, lightweight and merely opinions of important (or unimportant) people taken from media. We all have opinions on a variety of subject matter and we may enjoy expressing these anecdotally in social circles. But in academic and business terms, decisions must be made on the evidence. This evidence must be 'weighed' carefully. Decision makers must make sure that they are knowledgeable in a given subject area; they must not be ignorant. You must consider the weight of each piece of evidence, and place it on one side of the balancing scale; add your own evidence. The issue then becomes one of making judgements, very importantly not giving opinions or expressing personal views. The judgement should be made impartially, recognising the weight of the evidence on each side of the scale. Opinions, taken individually, are lightweight, anecdotal and prone to change. You may put yourself in the position of judge in a court; on the balance of the evidence before you, which at this stage of the research project is mostly the literature and the writer's experience, it is likely that 'xyz'.

The concept of weight is important; the analogy may be that a government-sponsored report executed by a team of leading people in the field may weigh 100 kg, whilst your own experience may weigh less than 1 kg. It is for you as the writer to judge the validity of each source and 'weigh' it accordingly. Do not use the weighing analogy too rigidly, since life is complex. At one point in time, given particular circumstances, the scales may drop most heavily on side 'A' and then at another point in time with slightly different circumstances, the scale may drop on side 'B'.

You must not be naïve as a writer. You must consider the validity of sources and take account of them when you weigh the literature. For example, a manufacturer will argue that its product is best. Bias and the motives for bias, e.g. money, job preservation, politics are powerful influences. The political world employs specialist spin-doctors. It is your job as the writer to skilfully weigh the validity of each source; be suspicious. Scientists have said that they never believe anything if it is written down; they have to validate it in the laboratory to believe it.

It is important to distinguish between isolated opinions and collective opinions sourced as part of the main research data collection exercise. Research places little emphasis on isolated opinions, especially if they are expressed spontaneously with little thought or reflection. Similarly, research is not interested in *your* opinion. Research is very interested to learn from the opinions of a few people during in-depth reflective interviews; these may be analysed qualitatively. Also, research wants to learn from the opinions of many people collected perhaps, from a survey for quantitative analysis. These opinions may form part of the main data analysis used to test, if appropriate, hypotheses. In a business sense, the collective opinions of people about products and services are drivers for buyer choice; clearly these

**Figure 3.1** Weighing the theory and the literature to make judgements (Microsoft, 2007).

opinions are very important. Since research does not want your opinion, do not let this leave you feeling 'worthless' in the process. The research exercise that you are undertaking is you; it 'smells' of you. It is your articulation of the problem, your use of the literature, your objective, your choice of method to meet the objective. The research requires you to use many of your skills and qualities to produce valid conclusions to an important problem. You will not express your opinion; you will be reflective, use insight, make interpretations. Based upon the evidence, you will make your judgements, as the expert, from a position of knowledge.

## 3.3 Sources of data

Some literature is more rigorous than others. The literature review needs to draw on the best evidence. Where a source is published is an indicator of its worth. Papers in academic journals are rigorously refereed, and are based on robust research methodologies. The refereeing process ensures that only the best research is published; attrition rates vary, but perhaps only one in five or one in 10 papers submitted are published. Authors are specialists in their field, and papers are often the output of funded research. They may be considered to be in the top division of written work. Within this top division, there are some journals which may be informally ranked in the upper quartile, some in the middle and some in the lower quartile. It is an accolade for academics to have their work published in the leading journal in their field. The analytical framework in the middle of some of these papers can be extremely complex; in these cases it is perhaps just sufficient that you digest the early parts and then the findings/conclusions at the end.

Conference proceedings are often useful sources of data. Some conferences are more prestigious than others. Conferences are not thought to be as prestigious as refereed academic journals, since it is more difficult to achieve publication in the latter. Other dissertations and theses can provide useful ideas about document structure and research methodology. They may provide a useful list of references. Reports commissioned by governments or government quangos are often very important. They may represent a turning point in the life of a discipline. Such reports are often well resourced, compiled by leading people in the field and therefore very authoritative. Similarly, reports commissioned by learned professional bodies or trade bodies can be useful. Textbooks often give leads to, or appraise, other authoritative literature. Some textbooks are written for practitioners, whilst others are for students to merely explain how systems *et al.* work. Student textbooks are important to bring your own base knowledge to the required level. The most critical material in textbooks may be where one or two people have acted as editors to bring together (though usually in separate chapters) the most authoritative writers in a field.

Weekly and monthly magazines in the discipline must be located and regularly reviewed; they are absolutely essential reading. These are published commercially and by professional bodies. The articles in such papers are sometimes merely gossip, items of news and the like; this is fine. They may, therefore, be considered in some parts to be lightweight and anecdotal. However, they often contain leads to important and current work elsewhere, e.g. government reports that are out for consultation. At the early stages of research projects they clearly give clues to what the issues of the day are, and where problems exist. You must also be mindful of the availability of audio and audio-visual material. Such material is often supported by written notes.

A literature review that is based on academic journals, conference papers and authoritative reports (flavoured by a sprinkling of textbook and magazine citations) is likely to score a much higher mark than one that is only based on textbooks and magazines. You need to demonstrate that the quality of your reading is at a higher level.

## 3.4 Methods of finding the literature

The expectations of a literature review may have increased in recent years because the process is assisted by electronic searching techniques. You must clearly endeavour to make best use of the Internet, commercial CD-ROMs, and library search catalogues. You should find out if there is a university that specialises in the topic area you have chosen. It follows that there may be several dissertations or theses at this university that you may be able to access electronically. Direct access is available, whether or not you are a student at this particular university, to library search catalogues.

Manually 'browsing' library shelves is equally as important as the electronic search. You must not think that all important information in a field is listed electronically. If the discipline relevant to the research project is engineering, it is useful to look for titles beginning with 'e' and 'j', viz: the engineering journal or the journal of engineering. Several years' editions of a leading journal in a field may be quickly browsed by looking at the index of articles on the front or back cover of the text.

The depth in a literature review is probably best obtained by gaining 'references from references'. Early leads can be obtained by electronic searching and browsing. Once an article is found, at the end of that article is often a list of references to other work in the field. When you obtain these references, you may find more references at the end of the new reference; and so the process goes on. Often such articles may not be available in the local library, so use has to be made of inter-library loan processes. The important point here is that it can be a time-consuming process, and so it is important to start the literature review early. Start the 'browsing' process well in advance of the allocated period of study.

At the end of the literature review process you should clearly know: (a) what the leading academic journals are; (b) what the lead industry magazines are; (c) what the leading conferences are; (d) where the centres of knowledge/excellence in the field are; (e) who are the relevant professional bodies; (f) which is the lead government department in the field; (g) which are the relevant government bodies/quangos; (h) which other countries in the world have an interest in this field; (i) who the national and international leading figures in the field are; and (j) what are the leading web sites in the field.

As well as your own university library, you may also use your local public library or another university library. The latter may be more convenient for travelling, or may contain a specific publication that you need to browse. Often universities have reciprocal arrangements whereby students from one institution may visit the library in another. Government agencies often have libraries related to business, and the environment. Joining a professional body is important. The UK professional bodies will argue (quite rightly), that they are the leading experts in their field in the world—quite literally the world. At the very least, they have access through their libraries to the most authoritative work in the world. A lot of the information is available electronically or can be posted to you. It may be necessary to be a student member to obtain material. If you are not a member, it is advisable that you join. Membership may be free to students; the time in your career that you may make most use of services provided by professional bodies is as a student. Many professional people, at least once in their lives, visit the headquarters of their professional body. If you can, go and browse the library shelves whilst you are a student.

## 3.5 Embedding theory in dissertations

Clough and Nutbrown (2007, pp.104–105) quote Lewin 'there is nothing so practical as a theory' and Silverman 'without a theory, there is nothing to research'. Kerlinger in Cresswell (2003, p.120) defined theory as 'a set of interrelated constructs (variables), definitions, and propositions that presents a systematic view of phenomena by specifying relations among variables, with the purpose of explaining natural phenomena'. Examples of flagship theories are theories of gravity first developed by Isaac Newton in 1687, Darwin's theory of evolution by natural selection in 1865, and Einstein's general theory of relativity in 1916.

Theory answers the 'why' and 'how' questions. Answers to 'why' questions often commence with 'because . . .'. It is often the case that answers to 'why' and 'how' cannot be seen with the naked eye. Explanations underpinning 'why' and 'how' may be extremely complex, and it is not reasonable to expect that they can be understood by all. Laws and principles arise from theories; they can only be written when theories are proven over time. If the theory is proven absolutely, a law can be written, such that 'if this happens, that will definitely happen', e.g. Boyle's law, 1662, is that 'the volume of a given mass of gas at a constant temperature is inversely proportional to its pressure'. Principles are written as guides, e.g. Archimedes' principle, (287–212 BC), 'any object, wholly or partly immersed in a fluid, is buoyed up by a force equal to the weight of the fluid displaced by the object'. People who do not understand the 'why' and 'how' explanations, may be happy to accept laws or principles without further question. Laws and principles make it possible to apply theories to practice without having to understand the 'why'.

The analogy of a spider's web can be used to illustrate the development of theories. Theories are meshed into the complex web of knowledge. Nodes on the web are variables, and strands between nodes indicate relationships between variables; cause and effect. As new research is executed, it may give strength by adding one new strand to the web, or by increasing the gauge of an existing strand. Alternatively, new research may break a strand. The broken strand may have been placed in position based on evidence available at that time, but the new research provides insight to discredit earlier beliefs. Strands that are placed in or removed from the web are not usually accompanied by grand announcements, since each individual strand is developed over a long period of time, and tested using multiple methodologies. Theories evolve and there are few 'Eureka' moments. Theories indicate the strength of any relationships and direction. The strength of relationships may be strong, such that movement of independent variables (IVs) will instigate similar movement in dependent variables (DVs); or not so strong, whereby large movement in IVs is followed by only small movement in DVs. Direction may be positive or negative; manipulating IVs up may cause DVs to go up, or manipulating IVs up may cause DVs to go down. Rather than the analogy of a web, in construction terms, you may wish to think of your work as merely putting a brick in the huge wall of knowledge, or more modestly, pointing or re-pointing mortar joints in that wall. Your dissertation may comprise testing of one very small part of a theory, using only one methodology.

Theories in hard sciences are proven to hold, both in laboratories and in practice, using many methodologies. Soft or social sciences involve human behaviour. In these fields there are lots of theories that have been found to hold, but they often have caveats in that a particular theory may apply given a certain set of circumstances, but if circumstances change, that theory may not apply. Most theories about people behaviour originate from disciplines outside construction. However, a construction context is sometimes used by researchers from other disciplines to conduct their work. There is a need to know whether theories developed elsewhere apply to construction. Researchers in construction therefore take theories from other disciplines and test them in a construction setting. In soft sciences, theories may compete, conflict and overlap with each other. For example, there are three competing leadership theories; trait, style and contingency.

Some students may start their study with examination of a theory. Qualitative studies may develop hypotheses or theories at the end, and include recommendations for testing by quantitative methods. The theory is the bedrock or foundation for the study, and it should be expressed at the start of a quantitative study. Construction students often start to describe problems from a very pragmatic or practical position. This may be particularly so if part-time students investigate industry-based problems. Research takes place to develop knowledge of society at large, and embed that knowledge such that it is a platform for application and for more research. It is therefore necessary to take descriptions of industry-based problems back to the theoretical foundations from which they originate.

To illustrate the gap between theory and literature, let them be placed on a continuum, with a numerical scale of 0–100. Let the lowest score, zero, for ease of illustration, be mere literature and the highest 100, theory. Let applied and pure research also be placed on this same scale, with applied research (or industry practice) scored at 0, and pure research scored at 100. Literature is not theory; it does not have that status. Items of industry news that appear in the press can be classified as literature, and their value is 0 on this scale, with no real value in research terms. As the academic quality of literature improves, it may have a higher rating, such that for example a scholarly reflective article may be scored at 10. Clearly, these numbers are not real numbers, but continuing the analogy, papers in refereed journals perhaps score 20. The journey required to classify something at the level of theory is a long one continuing over decades, and on this scale finishes, if it ever finishes, at 100. If you start your description of the problem with a theory in position 100, you can stay in that position if you intend your research to be 'pure', although there is potential for more marks if you can demonstrate application. If you start your description of the problem from a practically-based problem in industry in position zero, you should move along the continuum, and towards 100, to find a theory onto which you can 'hang' your study.

You should not find a theory just for the sake of finding a theory; your work should, in a very modest way, contribute to knowledge formation. Be assertive in your search for theories, using abstracts in electronic databases such as SCOPUS. This may involve exploration of literature in other disciplines, and will be part of your literature review. A thorough search should leave you overwhelmed with theories related to the problem that you have defined, and whilst you may review many of those as part of the literature review, select just one key theory for your investigation. The research question will narrow the number of possible theories.

Bothamley (1993) edited the multi-disciplinary text, *Dictionary of Theories; More than 5000 Theories, Laws, and Hypotheses Described*. Labels ascribed to theories are not always expressed with clarity. Some theories are mentioned more often in the literature than others, for example motivation theories or organisational theories. Some theories are mentioned less often, sitting quietly in the background known only to experts in that field. Some theories have their labels and are ascribed to the originator's name, e.g. McGregor's Theory X, Theory Y, whilst others are labelled according to their meaning, e.g. competition theory, economic theory (e.g. supply and demand), decision making theory, theory of planned behaviour, theory of design.

Some literature or authors use the word 'theory' loosely, claiming mere hypotheses to be theories, such as 'my theory is that quantity surveying is excellent training for a career in project management'. This is a hypothesis worth testing, but it does not have the status of theory established by testing and re-testing using different methodologies over time. Anecdotally, the media often reports conspiracy theories. These too are not theories, but more like hypotheses that are unproven. The relevance of some theories may decrease in modern society; some theories are cast in stone for ever, such as classical theories of organisations now over a century old, emanating from work by Max Weber in 1904. Newer theories, such as those in collaborative working, arguably have their origins over a mere few decades, and they are still under development.

It is possible to take a problem in practice or industry and base research on theoretical principles underpinning methodologies, rather than theory underpinning the problem itself. There may be debates about the validity of certain methodologies to test hypotheses, and their appropriateness to given problems. These debates may be around populations, sampling techniques, data collection, or analytical methods.

It is possible that in quantitative work your hypothesis will not be proven; no relationship is found between variables. You must not view this as though your study has failed. It cannot be that all hypotheses, expressed as provisional suppositions, are proven. The nature of research is that provisional suppositions are put forward as hypotheses for testing, but a positive outcome is not assured. If no relationship is found, this is fine; in the next study just back track a little and put forward new hypotheses for testing—'well if it is not A that causes B, I wonder if the cause is C?'

Consider three examples of taking industry-based problems back to theory:

(1) Site waste management plans (SWMPs) were made compulsory in 2008 for all construction projects over £0.3M. Contractors' problems include administrative and

training costs, and uncertainty about whether plans are effective and add value. Site managers complain about the administrative burden passed to them with no additional resource to help. The research questions are: (1) do SWMPs reduce waste? and (2) do SWMPs reduce cost? As an IV the quality of SWMPs can be expressed on a scale of 0 to 10 with zero being no plan, and 10 being a plan and its resulting actions arising from being 'best in class'. There are two DVs: waste and cost. Waste may be measured by the number of skips to tip or such like. Cost may be a calculation of staff training, time, extra cost of labour sorting and skips, etc. If relationships are proven, quantitatively, and in a positive sense, it would be found that SWMPs do reduce waste and cost.

There are theoretical issues to explore around the decision by Government to legislate for SWMPs, when in many spheres self-regulation or voluntary codes of practice are preferred. There are issues in social sciences about the propensity of companies and individuals to do what is right when faced with voluntary codes or legislation, and also the role of the state in protecting the environment. Perhaps the most appropriate theory is systems theory. SWMPs are systematic of government, local authorities (as enforcers), contractors, sub-contractors, suppliers and employees coming together to reduce waste. Systems theory 'predicts that the complexity of organisations, and therefore the role of management, will probably continue to increase—at least for so long as the efficiency-enhancing potential of complexity can continue to outweigh its inevitably increased transaction costs' (Charlton and Andras 2003). The research question could be that with the introduction of SWMPs 'has the efficiency-enhancing potential of complexity continued to outweigh increased transaction costs of SWMPs?' Also applicable may be the invisible hand theory developed by Adam Smith (1723–1790) quoted by Joyce (2001), '. . .being the managers of other people's money than of their own, it cannot well be expected that they should watch over it with the same anxious vigilance with which partners in a private co-partnery frequently watch over their own. Like the stewards of a rich man, they . . . consider attention to small matters as not for their master's honour and very easily give themselves a dispensation from having it'. Is it the case that construction companies delegate control of materials to others, who are motivated not to use those materials prudently, but to maximise their own earnings?

(2) In research involving materials testing, relevant theories include those around the chemical behaviour of constituent parts. These theories should be examined in the literature review. In tests of concrete, constituent parts can be varied by volume or by type. If new concrete mixes are tested, the results should be evaluated against theory. If a new mix has greater compressive strength, that is fine, but 'why'? The answer lies in the theory. Discussion around possible 'whys' should take place in the closing stage of dissertations.

(3) You work for a local authority that is reflecting whether to remove speed cameras. Cameras have been in place for a number of years. Accident data are available for periods before and after installation. Articulation of the problem revolves public and motorists' opinions and attitudes towards cameras, but the lead issue is whether they reduce the number of accidents. An initial approach may be to analyse the data using mathematical tools, and look for statistically significant differences in the number of accidents before and after installation of cameras. There is the possibility that findings may indicate either way; cameras do or do not reduce the number of accidents. There are theories in

the mathematical and statistical tools that you may challenge. There are also theories about the law being used as a deterrent to human behaviour.

To find theories, there is no substitute for general reading and browsing literature. If you are able to astutely link your work into the theoretical web, you give yourself the possibility of gaining many more marks. As an alternative to identifying a theoretical foundation for your study, you may place your work in the 'body of knowledge establishing best practice' (Seymour *et al.* 1997). There have been significant resources devoted to establishing best practice in many spheres of construction activity since Latham (1994), e.g. safety, procurement, culture. It is currently coordinated by Construction Excellence in the Built Environment. Whilst this may not gain marks so high as theory-founded research, it does place your work within a structured framework of applied research. It is implicit that best practice draws on theory. If you have not used the theory, it is arguable that it is only the case that you have not made the link yourself, and your marks should not be substantially less than studies that do. Without a base in either theory or best practice, your work is likely to make little contribution and score fewer marks.

## 3.6 Referencing as evidence of reading

Referencing is very important. Readers should be able to distinguish clearly between what is your original work and what is the work of others. You must use the work of others, and ensure that it is cited. You should stand on the shoulders of others, not try to reinvent the wheel. A valid research project may be to take the published data of one researcher and analyse it in a different sort of way—provided that the original source of the data is acknowledged.

Referencing is not just about getting the technical details of referencing correct, although that is important. It is the tool that you use to demonstrate to examiners and other readers that:

- You have been reading
- You are not passing the work of others off as your own
- You can digest and appraise the work of others in a critical sort of way

If you do not show evidence of reading in your work, you cannot pass. If you pass the work of others off as your own, you will be failed (plagiarism). If you do not critically appraise the work of others, you are likely to get lower marks. Being able to reference in a technically correct way is therefore essential. Some students may conscientiously read extensively and merely not cite their sources. This is not good enough. Using the analogy of courts, judges cannot reach decisions based upon material presented to them that may (or may not) be spurious.

Some students have really good practical experience in some topic areas. Clearly they are in a strong position, and can bring the weight of their experience into arguments being made. However, to write an academic piece of work purely based on practical experience is not good enough either. After all, a student is unable to write authoritatively about a topic area, without having digested the evidence collated from the experience and research of others. Students with practical experience have a strong platform from which to build, but they must integrate that experience with academic tools; that is, they must read, and in their writing blend their practical experience with other evidence in the literature.

## 3.7 Citing literature sources in the narrative of your work

There are two separate components to referencing: citing literature sources in the narrative of your work and then giving the full reference details in a separate 'section' towards the end of your document.

The two components briefly illustrated are:

(1)  The name of the author and year in the narrative of your work, e.g. Taylor (1911) stated that . . .
(2)  In the 'references' section, provide precise details of where this work can be found: Taylor, F.W. (1911) *Principles of Scientific Management*. New York and London: Harper & Brothers

These full reference details are in a 'section' and not a 'chapter'. The last chapter is usually conclusions/recommendations with a number, say Chapter 6. References follow after the conclusion without a number, then the bibliography and then appendices.

If readers, after reading the narrative, are motivated to seek out a source you have cited, they can turn to the reference section. The references will be in alphabetical order. Thus if the text by Taylor (1911) is of interest, the reader can turn to the references section to learn the full reference details.

Literature may be cited verbatim or paraphrased. Verbatim is word for word. If you do cite verbatim, you must put that section of text in speech marks; it is speech marks that distinguish verbatim text from paraphrasing. If you cite verbatim, be careful not to do it out of context. Do not repetitively use long sections of text from the literature. Verbatim quotes may most often be just sentences or parts of sentences. The source of the quote may be written as part of the flow of a sentence with only the year given inside the bracket, thus:

Ruskin (1860) argues that '. . . it is unwise to pay too much, but it's worse to pay too little'.

Alternatively, it is not written as though part of the flow of a sentence, with the author and year given inside the bracket, thus:

' . . . it is unwise to pay too much, but it's worse to pay too little' (Ruskin 1860).

Both methods are acceptable. To make your writing as prosaic as possible, perhaps use both methods interchangeably as you cite different authors.

It is not appropriate that a literature review will have lots of large paragraphs of text taken from other sources verbatim. If there is a whole paragraph that is central to an important argument being made around an objective, then by all means cite it all. When students do cite paragraphs of text verbatim, it is often without speech marks, and the source is merely indicated at the end of the last sentence. For the reader it may not be possible to distinguish how many sentences are verbatim, and a charge of plagiarism may arise. Therefore, for clarity, it is convention that these long sections of verbatim text, as well as being in speech marks should also be set apart from the main text, in italics (although italics contravenes the disability literature) and indented, thus:

> *'It is unwise to pay too much, but it's worse to pay too little. When you pay too much, you lose a little money—that is all. When you pay too little, you sometimes lose everything, because the thing you bought was incapable of doing the thing it was bought to do. The common law of business balance prohibits paying a little and getting a lot—it can't be done. If you deal with the lowest bidder, it is well to add something for the risk you run. And if you do that, you will have enough to pay for something better'* (Ruskin 1860)

To paraphrase is to read a section of text, whether that be a paragraph, page, chapter or whole book, then to digest and understand what the writer is saying, and to summarise it more succinctly in your own words. The words that you choose must not misrepresent what the original writer has said, either accidentally or otherwise. Misrepresenting the work of others can lead to the 'Chinese whispers' effect. As work passes from writer 'A' to writer 'B' to writer 'C', etc., what comes out from the last writer is radically different to what was written by the first. You must execute any paraphrasing accurately and fairly. The paragraph from Ruskin (1860) may be (arguably accurately) paraphrased as:

> Ruskin (1860) articulates the argument that seeking the lowest price may leave you with high risks, and ultimately higher costs.

Note there are no speech marks since these precise words did not come from the pen of Ruskin.

## 3.8 Secondary citing

Secondary citing is used when you read what one author 'says' another author has 'said', and then wish to cite the first author based on what the second author has written. If we are being suspicious, we could say what the second author alleges the first author has said. The need for secondary citing may occur often. Secondary citing should be a signal for readers to be careful. It is life that all forms of communication are at risk of being misinterpreted as they pass from one source to another. It is part of your task to make sure that you do not contribute to any misinterpretation. It is probably not a good idea to paraphrase something that has been paraphrased.

An example of the need for secondary citing may occur in management studies, if you wish to cite the work of F.W. Taylor. He was born in 1856 and died in 1915. He is known as the father of scientific management. One of his most famous texts was:

> Taylor, F.W. (1911) *Principles of Scientific Management*. New York and London: Harper & Brothers

It is highly unlikely that you will have read the 1911 text. Unless your whole study was around his work, there is not a reasonable expectation that you will read it. Learning of Taylor's work through secondary sources is fine. If you have not read the text, it is important not to suggest that you have. To secondary cite, therefore, just write 'Taylor (quoted in Smith 2010) states that . . .' You should not give the full reference details of Taylor's text in your work; that may mislead examiners to think you have read it first hand. Readers who may wish to follow up the 1911 reference to Taylor must do so through Smith (2010).

As another example, you often need to refer to authoritative construction industry reports that have been published. They might be very difficult to locate. If they are in professional body library archives, since they are valuable and out of publication, they may be only available to read in person, not for loan. Examples may be the landmark reports by Simon (1944) and Banwell (1964). A text such as Murray and Langford (2003), referenced below, provides a compilation *et al.* of the reports, and it would be a perfectly acceptable secondary source to cite:

Banwell, H. (1964) The Placing and Management of Contracts for Building and Civil Engineering Work. London: HMSO
Murray, M. and Langford, D. (Eds) (2003) *Construction Reports 1944–98*. Oxford:, Blackwell Science
Simon, E. (1944) The Placing and Management of Building Contracts. *The Simon Committee report*. London: HMSO

## 3.9 Who to cite in your narrative

Citations should be to the author. If there are two authors cite them both, e.g. Smith and Brown (1900) argue that . . . . or . . . (Smith and Brown 1900). If there are three or more authors, e.g. Smith, Brown and Baker, cite it in your text as Smith *et al.* In your reference section towards the end of your document it would give all three authors' names, e.g. Smith, A., Brown, B. and Baker, C. (1900). '*Et al.*' is Latin for 'and others'; foreign phrases by custom should be in italics.

If the name of the author is not given, cite the organisation, e.g. BBC (2010) or the source, e.g. *The Times* (2010). If it is not possible to locate any of these, as a last resort use anon (anonymous), e.g. anon (2010).

If authors have more than one publication in a year that you wish to cite, the first one in your document should be given the letter 'a', then 'b' and so on, e.g. Smith (2010a), Smith (2010b). These letters should appear in the narrative and in the reference section.

It is important to point readers directly to a source with its page number. Readers may often wish to follow up a citation and read more about the surrounding context in the original work. It can be very frustrating to be directed to a textbook only, and then not to be able to locate relevant parts of the text. There is an issue about whether to include page numbers in the narrative part of the text or in the reference section at the end. As a general rule it is best to keep as much information out of the narrative as possible so that readers are not distracted from the flow of the text. On this basis, put the page numbers in the reference section. However, if a source is cited more than once, it is necessary to give the page numbers at all points cited in the narrative, e.g. (Smith 2010, p.32), (Smith 2010, p.105). The details of the Smith publication is then written out in full only once in the reference section, without page numbers.

It may be the case that you may wish to cite the generic work of previous writers. If the writer has been prolific, this might just be by giving the author's surname, and no references at all to specific texts in the reference section. Alternatively, if the writer has just one key text, referring to that text with its year in the narrative, e.g. Smith (2010), then include the text in the reference section.

If you were citing not just 'any' Smith (2010), but someone who is extremely authoritative, you may wish to emphasise this by saying something like 'the chief executive of ABC construction, John Smith (2010) argues that . . .'. To emphasise that this is not just 'any Taylor', nor just any of Taylor's books, you may write something like: F.W Taylor, the father of scientific management, in his seminal text *Principles of Scientific Management* in 1911, stated that '. . ..'. (Smith 2010). Also, rather than just cite HSE (2005), you may wish to state 'the HSE publication *Health and Safety Induction for Smaller Construction Companies* (HSE 2005) states that '. . .'.

## 3.10 References or bibliography or both?

Definitions of bibliography and references vary. The two words are used interchangeably, although they are different—or are they? Definitions may vary between academic disciplines, universities, university departments or individual academics in the same department. The two key authoritative sources for you must be your university department and supervisor. Tell supervisors if their definitions conflict with departmental definitions, and seek clarification. The definitions used here are:

- References—everything cited in your dissertation
- Bibliography—everything that you have read or browsed that is relevant to your subject area, but has not been cited

Some authors use a broader definition for a bibliography, to include all material relevant to a subject area, even though the writer has not read or browsed it—perhaps this is not appropriate for your dissertation.

References may be considered the most important; they are what you have read, digested, understood and used in the appraisal of your subject area. The fact that literature has been 'cited' is the 'evidence' that you have done the reading. If you cite a book, it is not expected that you have read the whole of the book. You may have browsed the whole book, and read relevant parts. Cite the parts you have read and digested. If you are citing one page, cite it singularly as say 'p.10'. If it is a range of pages, and you will paraphrase what has been written, cite it in the plural as say 'pp.10–15'. Marks are awarded in dissertations for evidence of reading and for the way you have used the literature in your work; evidence for this comes partly from references. Examiners will 'always' browse the list of references at the end of your work to make a judgement on the extent and the academic weight of your reading; remember weight comes from the type of material you have cited. Reading and citing weekly construction magazines is to be encouraged, but this will not carry so much weight if it is done at the exclusion of academic papers and other authoritative sources.

A good bibliography may serve two purposes. Firstly, students may wish to demonstrate to examiners that they have read more widely than just the references, and therefore get credit or marks for that reading. It may be that having done the reading, the material was not contentious, not directly relevant to the objectives, or not relevant to a wider argument.

Secondly, it might be that subsequent researchers may be interested in bibliographies, as they look for leads from other literature.

Examiners may be suspicious of bibliographies. It is possible that having completed a dissertation, students merely locate textbook title details electronically and pass it off as though 'read' in a bibliography. This cannot be done so easily if using references, since within the flow of the narrative work the text will be cited, e.g. you cannot write in your narrative 'Smith (2010, p.10) states xyz', if you have not had that text by Smith open at page 10. Perhaps a sensible position is that if you have done little more than casually browse some literature, do not include it in your bibliography. If you have read a chapter or so, do put it in and include in the bibliography the pages you have read. You may have a viva at the end of your process; you do not want to be embarrassed if you cannot speak with sincerity about your bibliography.

So, to answer the question 'references or bibliography or both?'; ideally both. Some examiners may not be worried at all if there is no bibliography. However, if there are no references, the dissertation will fail.

## 3.11 Common mistakes by students

In the narrative, do not:

- Give abbreviated or full web addresses: golden rule – no 'www' or 'co.uk' or 'org.uk' or 'com' anywhere in the narrative. Do include these web addresses under the heading 'references' at the end of the document
- Give the title of the text unless you want to emphasise the stature of a particular piece of work
- Detach the information from its source by having a full stop in the wrong position. This is the correct way: 'It is unwise to pay too much, but it's worse to pay too little' (Ruskin 1860). The following is incorrect, since the full stop should be after the close bracket: 'It is unwise to pay too much, but it's worse to pay too little'. (Ruskin 1860)

In the reference section, do not:

- Use the abbreviated '&'
- Number references; you must put them in alphabetical order of author
- Bullet point references
- Arrange references under separate headings of textbooks, journals, reports and web pages
- Use the combined title of references and bibliography; you must keep them separate
- Put references at the end of each chapter in the dissertation; they should be inserted towards the end of the document in one section
- Repeat sources in both the reference section and the bibliography

Citing web pages often causes students difficulties. In bibliographies students may cite a generic web site for an organisation they have been browsing, e.g. http://www.constructingexcellence.org.uk/. This is not appropriate; whether it is a reference or in the bibliography, it must be to a specific web page. References are often given starting with the web address http://www. etc. This is not correct. Always start with the surname of the author or the organisation first, e.g.

> Constructing Excellence (2009) 'G4C Egan Report' http://www.constructingexcellence.org.uk/news/article.jsp?id=9865 [09 September 2009]

The issue about the use of abbreviations and acronyms arises both when referencing and in your narrative text. In your narrative, convention is that at the first point of citing in your dissertation it should be spelt out in full, with its abbreviation in brackets, thus: The Health and Safety Executive (HSE 2005) states that ... At subsequent citings it can then be just 'The HSE (2005) states ...'. If readers later 'forget' what HSE stands for, they can refer to a list of abbreviations at the front of your dissertation. In the reference section both the full and abbreviated name can be given thus:

> HSE (2005) *Health and Safety Induction for Smaller Construction Companies.* The Health and Safety Executive. http://www.hse.gov.uk/construction/induction.pdf [09 September 2009]

## 3.12 Using software to help with your references

You have a choice of three methods to compile your references at the end of your dissertation:

(a)  Do it 'manually'; that is use the word processor to type in all the information required
(b)  Use tools in word processing software, e.g. the references drop-down menu in Microsoft Word 2007 or similar, or
(c)  Use specialist web-based software such as End Note or RefWorks

In many aspects of learning it is necessary to grasp basic underpinning principles. Understanding may be best digested manually before using electronic or software aids; for example, it is better to learn mental arithmetic before using a calculator. Therefore, at least some time using method (a) would seem appropriate.

There are hundreds of different referencing styles. Some are based on British Standard BS 5605: 1990 *Recommendations for citing and referencing published material,* 2nd Edition. Others are based on the International Standards Organisation (ISO). Many are based on styles adopted by different academic journals in different professional disciplines. Two common styles are the Vancouver system and the Harvard system. A humanities coursework (e.g. literature, philosophy, history, art and design) normally uses the Vancouver system. Social sciences or technology (e.g. education, health, sociology, psychology, business, engineering) normally use the Harvard system. You must follow the guidance of your university department; in the built environment this is likely to be the Harvard system. There may be slightly different interpretations of the Harvard system. Follow your own university guide to the letter. This book is using an adaptation of the British Standard Harvard version.

If you are to record and manage your references 'manually', it can be difficult to closely follow the requirements of either Harvard or Vancouver systems with precision; if you decide to do it manually, keep practising and keep your university guide in your briefcase as though a dictionary. It may help if you set up an electronic file to record reference details of the work you have read as you read it. Begin writing up the references/bibliography at the start of the dissertation. It may be unclear at the start as to whether articles will be a reference or part of the bibliography, but at the writing-up stage they can be easily allocated to the appropriate section using word processor cut and paste functions.

You need to aim for perfection in writing references/bibliographies; full stops, semicolons, colons and commas should all be in the correct place. All information should be in the correct sequence, with no information missing, and consistently presented. Precision is required so that libraries can locate cited articles with ease, without having to ask for more details or clarification. If you were to ask for the 'Oxford' book, is that the name of the book, the author, the publisher or the place of publication? More importantly for you, if you need to go back to the library to relocate a text you have had previously, you need to be able locate it quickly, including relevant page numbers.

As an example, generically, the British Harvard System requires the following information in the reference section:

- Author's surname
- Publication year
- Title
- Edition
- Place of publication
- Publisher

One issue with software is the time required to learn to use it, and whether that time invested will save time later. Referencing is not just for dissertations, it is for academic work at all levels. If you invest your time to learn to use the software in modules completed before the dissertation starts, there are greater potential savings.

This is what the Microsoft Word processing software, as noted in (b) above, will do for you:

- Offer you a limited number (10) of different methods of referencing or 'output styles', e.g. APA
- Offer you the format required for a variety of sources (16), e.g. book, journal article, electronic source
- For each of those 16 sources, prompt you to type in relevant details. For a book this may be author, year, title, place of publication, publisher. For conference proceedings this may be author, year, title, editor, proceedings title, conference date, publisher, pages
- Hold all the references that you 'save' as a 'master list'
- Draw sources from your master list into your narrative; that is known as 'citations whilst you write'
- Compile and sort your references at the end of your work, including adding all punctuation

Specialist software is more sophisticated than word processing software. This is what the specialist software, noted in (c) above, will do for you:

- Everything that the word processing software does
- Offer you a wide range (hundreds) of different methods of referencing or 'output styles'

- Offer you the format required for more different sources (say 30), e.g. journal by paper, journal electronic
- Allow you to hold, copy and paste between folders; perhaps you may have folders for different assessments that you are doing
- Allow references from other electronic sources to be copied and pasted so that the data does not have to be manually typed into the software
- Import references from the software into your word processed document
- Allow you to access your references remotely, through the Internet

You have to make your decision; is it to be (a), (b) or (c)? If it is (a), this leaves you vulnerable to making mistakes, and having to spend more time, probably at a later stage of your study, making sure that your references are the best they can be. The most sensible option is clearly (c). To learn to use the software initially is just a matter of an hour or two. Expertise will develop as you use it more, and it will take far less of your time in use than method (a). The key saving features are:

- The dialogue boxes act as 'prompt' facilities—otherwise it is more difficult to distinguish between what types of material you need to record for different sources
- Importing references from electronic sources—otherwise you have to type in data manually, and
- The sorting of references into order and insertion of punctuation at the end of your work—otherwise you have to do this manually

Library or departmental tutors may be willing to give demonstrations to groups of students. Alternatively, just browse through the help facilities alone. Table 3.1 is based on the RefWorks software and the British Standard Harvard version. It illustrates the referencing details required for 19 source types; that is journal by paper, journal electronic, etc. Also illustrated is the sequence in which all required details should be presented in the referencing section. There is much to remember manually; you need the support of the specialist software. It is further complicated by some systems requiring such things as the titles of whole works to be in italics, to be partly underlined or in uppercase letters. If you are doing it manually, this is perhaps fine; Table 3.1 may help you. However, by browsing the table, you will appreciate that manual methods are likely to be too time consuming, and you will be inspired to use one of the software packages.

## 3.13 Avoiding the charge of plagiarism

Plagiarism is defined as 'take and use (the thoughts, writings, inventions, etc. of another person) as one's own'. There are sophisticated and powerful software tools that universities are routinely using that 'ensure originality as well as use of proper citation' (Turnitin 2009); in other words, to spot instances of plagiarism or cheating. You may be asked as a matter of routine to submit your dissertation electronically, or if you have submitted only a paper copy of your document, the university may later ask you for an electronic version, if it has some suspicions. The software is a vast repository of data and holds an electronic record of 'everything that has ever been written'; not literally of course, but with the passage of time it

**Table 3.1** Referencing details required for 19 source types (numbers indicate the sequence in which information should be given).

| Rows = type of information required. Columns = information sources | Generic | Web page | Book whole | Book edited | Conference proceedings | Journal article | Journal electronic | Magazine article | Newspaper article | Personal communication | Report | Computer programme | Case/court decisions | Laws/statutes | Dissertation/thesis | Dissertation/thesis unpublished | Online discussion forum/bloggs | Video/DVD | Unpublished material |
|---|---|---|---|---|---|---|---|---|---|---|---|---|---|---|---|---|---|---|---|
| Author/s | 1 | 1 | 1 |  | 1 | 1 | 1 | 1 | 1 | 1 | 1 |  |  |  | 1 | 1 | 1 |  | 1 |
| Publication year | 2 |  | 2 | 2 | 2 and 7 | 2 |  | 2 | 2 | 2 | 2 | 2 | 2 | 2 | 2 | 2 | 2 | 2 | 2 |
| Title | 3 | 4 | 3 | 3 | 3 | 3 | 3 | 3 |  | 3 | 3 | 3 | 3 | 3 | 3 | 3 | 3 | 3 | 3 |
| Edition | 4 |  | 4 | 4 |  |  |  |  |  | 4 |  | 4 |  |  |  |  | 4 | 4 |  |
| Place of publication | 5 | 5 | 5 | 5 | 8 |  | 5 |  |  | 5 | 5 | 5 |  | 5 |  | 5 | 5 | 5 | 5 |
| Publisher | 6 |  | 6 | 6 |  |  | 6 |  |  | 6 | 6 | 6 | 6 |  |  |  | 6 |  | 6 |
| Last updated - full date |  | 2 |  |  |  |  |  |  |  |  |  |  |  |  |  |  |  |  |  |
| Last updated - year |  | 3 |  |  |  |  | 2 |  |  |  |  |  |  |  |  |  |  |  |  |
| URL |  | 6 |  |  |  |  | 4 |  |  |  |  |  |  |  |  |  |  |  |  |
| Accessed year |  | 7 |  |  |  |  |  |  |  |  |  |  |  |  |  |  |  |  |  |
| Accessed, month/day |  | 8 |  |  |  |  |  |  |  |  |  |  |  |  |  |  |  |  |  |
| Editor/s |  |  |  | 1 | 4 |  |  |  |  |  |  |  |  |  |  |  |  |  |  |
| Proceedings title |  |  |  |  | 5 |  |  |  |  |  |  |  |  |  |  |  |  |  |  |
| Volume |  |  |  |  |  | 5 |  | 5 |  |  |  |  |  |  |  |  |  |  |  |
| Periodical, full |  |  |  |  |  | 4 |  | 4 | 4 |  |  |  |  |  |  |  |  |  |  |
| Issue |  |  |  |  |  | 6 |  | 6 |  |  |  |  |  |  |  |  |  |  |  |
| Start page |  |  |  |  | 9 | 7 |  | 7 | 6 |  |  |  |  |  |  |  |  |  |  |
| Other pages |  |  |  |  | 10 | 8 |  | 8 | 7 |  |  |  |  |  |  |  |  |  |  |
| Conference date |  |  |  |  | 6 |  |  |  |  |  |  |  |  |  |  |  |  |  |  |
| Article title |  |  |  |  |  |  |  |  | 3 |  |  |  |  |  |  |  |  |  |  |
| Sections |  |  |  |  |  |  |  |  | 5 |  |  |  |  |  |  |  |  |  |  |

(continued)

**Table 3.1** (Continued)

| Rows = type of information required. Columns = information sources | Generic | Web page | Book whole | Book edited | Conference proceedings | Journal article | Journal electronic | Magazine article | Newspaper article | Personal communication | Report | Computer programme | Case/court decisions | Laws/statutes | Dissertation/thesis | Dissertation/thesis unpublished | Online discussion forum/blogs | Video/DVD | Unpublished material |
|---|---|---|---|---|---|---|---|---|---|---|---|---|---|---|---|---|---|---|---|
| Newspaper name | | | | | | | | | 4 | | | | | | | | | | |
| Report number | | | | | | | | | | | 4 | | | | | | | | |
| Developer | | | | | | | | | | | | 1 | | | | | | | |
| Counsel | | | | | | | | | | | | | 1 | | | | | | |
| Ordinal series | | | | | | | | | | | | | 4 | | | | | | |
| Jurisdiction | | | | | | | | | | | | | 6 | 6 | | | | | |
| Sponsor | | | | | | | | | | | | | | 1 | | | | | |
| Type | | | | | | | | | | | | | | 4 | | | | | |
| Institution | | | | | | | | | | | | | | | 4 | 6 | | | |
| Degree type | | | | | | | | | | | | | | | 4 | | | | |
| Distributor/studio | | | | | | | | | | | | | | | | | | 6 | |
| Director | | | | | | | | | | | | | | | | | | 1 | |
| Type of work | | | | | | | | | | | | | | | | | | | 4 |

comes closer to that. It may take the software a day or so to search your document for matches to information that it holds in its repository. The result of the software search is that it gives a result in percentage terms, e.g. 15% match. Anything above this figure may initiate some concerns. However, a 30% match may be perfectly acceptable, *providing* you have cited the source of the material in your work. An important point to be clear about here; if you have cited the source, you have correctly acknowledged that this work, or these words, are not your own. You are legitimately using the work of others to support a point you need to make. Indeed, you have to do this; you have to read and develop your knowledge and enhance society's knowledge from a platform created by others. However, if you do not cite the source, you are passing someone else's work off as your own—cheating. The penalties for plagiarism, particularly in later years of study, can be severe, including failure or even expulsion from university. Universities may retain a plagiarism register, so that repeat 'offenders' can be identified.

As universities are increasingly using such software, the expanse of repositories increases significantly. Work submitted by students becomes part of the repository. Therefore, student 'A' may cite an original piece of work that was never produced electronically. This may have been as far back as the 1800s, or more recently in the 1990s. When student 'A' submits an assessment, a reference to the 1800s or 1990s original work is 'logged'. A short time later student 'B' cites the same 1800s or 1990s work; the software will give a match for student 'B' to the work of student 'A'.

It is for you, and not others, to check your own work before submission. You should be able to submit a draft of your work before submission to your tutor, so that you can identify any potential problems yourself, such as missing citations. Do not view the software with fear. Your view should be that it is a fantastic tool that:

- Prompts you to insert citations in drafts of documents that have innocently been missed, and more importantly
- Reinforces underpinning concepts of academia, such as the necessity to read and bring the weight of the evidence of your reading into your writing

## Summary of this chapter

The literature review is essential to establish a baseline for the remaining parts of your study. It is likely to be a substantial part of the dissertation. At the end of the literature review you should be able to make judgements rather than give opinions, and convert provisional objectives to firm objectives. You should manually browse paper-based material extensively, as well as using electronic search tools. Academic journals are more rigorous than industry magazines, and you should try to incorporate both in your review. Similarly, theory is more rigorous than mere literature, and you should try to embed the theory in all parts of the dissertation. Citing sources, providing references and bibliographies should be done using the correct technical methods. Do not plagiarise in your work; if you do, you may fail.

# 4 Research goals and their measurement

The objectives of each section of this chapter are:

4.1 Introduction: to provide a context for defining research goals

4.2 Aim: to define and provide an example

4.3 Research questions: to define and provide an example

4.4 Objectives: to define and provide an example; to emphasise the importance of objectives

4.5 Objectives with only one variable: to provide examples

4.6 Objectives with two variables: to provide an example

4.7 Hypotheses: to define and provide examples, including the null, the alternative, the one tail and the two tail

4.8 Independent and dependent variables: to distinguish between the two and provide examples

4.9 Lots of variables at large, intervening variables: to define and provide examples

4.10 Subject variables: to define and provide examples

4.11 No relationship between the independent variable (IV) and the dependent variable (DV): to explain the consequences

4.12 Designing your own measurement scales: to illustrate examples that do and do not need your own design

4.13 Levels of measurement: to define categorical, ordinal, interval and ratio

4.14 Examples of categorical data in construction: to provide examples

4.15 Examples of ordinal data in construction: to provide examples

4.16 Examples of interval and ratio data in construction: to provide examples

4.17 Money as a variable: to identify how it may be measured

## 4.1 Introduction

Research goals may be used as the umbrella term to encompass all things that you will do in your dissertation; viz., aims, research questions, objectives, hypotheses. Definitions vary in the literature, particularly the definitions of aims and objectives.

*Writing a Built Environment Dissertation: Practical Guidance and Examples.* Peter Farrell.
©2011 Peter Farrell. Published 2011 by Blackwell Publishing Ltd.

Linking into all of these is the title that you will choose for your dissertation. The title needs to be attractive to entice potential readers, but also concise and accurate so that it does not mislead. If your dissertation is made available electronically, it is the title that is primary in attracting search engines. The title should be different to the aim, but may typically include a mix of terms from the aim and objectives. It may also give an indication of what was found or concluded in a study. A title set at the beginning of the process can be provisional.

Sometimes students do not write research questions; it is better that they do, since these questions assist in the logical progression of studies and illustrate understanding of how objectives and hypotheses are derived. The sequence should be that the articulation of the problem gives rise to research questions; the objective follows, as a statement of what you will do to answer the question. A hypothesis is finally written from the objective, in such a form that it is suitable for testing. Sometimes students write objectives that are similar to questions; similar is not good enough. If they are only similar, it follows that there are differences. Differences can be so distinct that if dissertations meet objectives, they do not answer the questions that originated from the problem. There should not be any differences; questions, objectives and hypotheses should imitate each other, with precision. It is suggested that dissertations are bound around objectives; this is because industry and academia may come together more easily when discussing objectives rather than questions or hypotheses. Therefore, when there is a need to refer to a research goal (question, objective or hypothesis), refer to the objective, except perhaps if you are to refer to a research goal before a statistical test; in this case refer to the hypothesis. It is permissible that dissertations are alternatively bound around questions or hypotheses; you decide. It just becomes a matter of semantics.

Some students may see the possibility of getting some really good data and therefore do the study 'back-to-front'; that is, they set the objective to suit the data that they can get. Whilst on the one hand this is arguably not the way to do industry-based research, getting data and having a really good 'poke-around' it, may give some important insights. Researchers may be asked at an early stage of the study what the proposed methodology will be. A legitimate answer for you could be 'I don't know, I will pick the most appropriate methodology to meet the objective and that objective is currently only provisional'. Altering the objective only slightly could radically alter any proposed methodology. However, it is useful to have a methodology in mind at an early stage of the study. If you are asked to present a proposal, you may label your methodology provisional.

## 4.2 Aim

The aim is the ultimate goal of the study; it is a statement at a strategic level. The introduction to the dissertation will often start by saying 'The aim of this study is to investigate. . .'. The aim will be a statement of what the dissertation will attempt to do—often in the form of what is to be investigated for qualitative work or what impact the IVs may have on DVs for quantitative work. The aim will identify the context of what is to be attempted; what field are you in? If the aim is carefully written at the outset, it should not change from beginning to end. If you reflect at some later stage of the process that the words used in the aim are not sufficiently succinct, by all means change them, but try and do it without altering in your own mind the strategic direction of your study. The aim will give you direction for your objectives, and if objectives change at the early stages of a study, they should remain within the remit of the aim. The aim can be written without consideration of resource limits. Resources will always limit researchers, particularly personal time, and therefore dissertations should not be judged on the yardstick of the aim. Aims will not be achieved. Small studies have one aim. There are analogies with business; mission statements of companies express aims. They are written at a strategic level, designed in boardrooms by directors or partners. Objectives are statements at an operational level. They support the mission statement in a business sense, or support the aim in the sense of a dissertation. Operational level is a level below the boardroom; that is, managers who are closer to the 'shop floor'.

Military history across the world can be used to illustrate the relationship between aims and objectives. The aim of a war may be to defeat an enemy, and this aim is supported by a series of military objectives that may involve battles on the one hand and intelligence on the other. If the objectives are not met, the aim cannot be achieved. In a construction context a study may start with an aim around customer satisfaction in the context that an industry is losing clients, and then as the study develops, one of the objectives may be around the quality of product.

## 4.3 Research questions

An early part of the dissertation would normally involve the articulation or description of the problem being investigated. Arising from the problem should be research questions. At this point the substance of the dissertation may stand or fall. A well-articulated problem will almost automatically 'spit out' questions. Independent readers of your narrative should be able to tell you what your questions should be. Your early exploratory work, perhaps speaking informally with professionals, could be around matching up the description of the problem with the questions and making sure that your perception of that match mirrors the perception of others. The research questions need to be strong, robust, relevant to the topic area, and in the context of industrial problems, addressing real issues of importance. Weak research questions will mean that the remainder of the dissertation is merely an academic exercise of data collection and analysis. A preference for two variables in objectives stems from two variables in questions, e.g. what effect will a move in interest rates (IV) have on inflation (DV)? Research questions should state the position for the argument or investigation—written in the form of 'what' or 'how'.

## 4.4 Objectives

Objectives must be capable of having an outcome, and the success of the dissertation will be measured against them; they are statements of a dissertation's hoped for outcomes. Objectives are often preceded by words such as 'to determine', 'to assess', 'to compare', 'to design', 'to determine', 'to develop', 'to establish', 'to evaluate', 'to examine', to find out', 'to measure', to review', 'to show', 'to survey', 'to test'. They are statements at an operational or tactical level. They take the aim and recognise constraints to translate the aim into 'do-able' statements—what the study hopes to achieve or discover.

Objectives should be written into the narrative as the introduction develops, so that the reader can determine exactly where the study is going; how the objectives will be met and what is the basis of that decision. Perhaps a useful strap-line is that:

'an objective is a statement of what you will do, and you will 'damn well' make sure you do it . . . lest you fail'

Perhaps it is also useful to write your objective as a strap-line, e.g.

'to determine the level of compliance of UK contractors with best practice in health and safety'

It would have been developed from, and would imitate the question, which was also written as a strap-line, 'what is the level of compliance of UK contractors with best practice in health and safety'? The strap-line is something you can carry round with you at the forefront of your mind, and when somebody asks you what you are doing for your dissertation, this is your verbatim response. In conversation with others you should then be able to explain the nature of the problem from which the objective was derived, and you may have in mind a provisional methodology to meet this objective. In your dissertation you will use carefully selected words to define variables in your objectives. The definition will form the basis of the research instrument that you use to measure the variables.

You should work long and hard at your objectives; labour over each word. They may be the result of many iterations. At the early stages of your study objectives may be considered provisional. Your objectives may change as you develop the early part of the study. Those changes may result from some early exploratory interviews or from new insights gained from the literature. At the end of the literature review, your objectives should become written in stone; it is then for you to go out and merely meet your objectives.

In many completed studies objectives 'wobble like jelly'. Objectives are not stated with clarity; if they are stated, they change as dissertations develop. The methods used, perhaps questionnaires, measure some concepts, often poorly, and are not related to the objectives. Sometimes objectives disappear part way through the document, and then re-emerge towards the end under a different label (e.g. reason for the study) with some key words changed or shuffled around. To avoid the wobble or slip of even one word, do not retype objectives as you need to repeat them; copy and paste them from the introduction.

The whole study must revolve around the objectives. At the risk of being monotonous in your dissertation, mention them in each chapter if necessary. The objectives must be in the introduction, and the literature review can be rounded-off by reaffirming the objectives. The description of the method could have a precursor something like 'this is the method used to meet the objective, thus: . . .'. The analysis and discussion chapters could similarly state 'this is the analysis (discussion) to meet the objective, thus . . .' In the conclusion the objective strap-lines may be given as headings.

Do not make readers work hard to find your objectives buried in the middle of large paragraphs of text. You may find it useful to number each of your objectives and to set them apart with a tab set, thus: The objectives of the study are:

(1)   To understand and appraise the UK planning system
(2)   To determine to what extent the UK planning system deters developers

Numbering objectives may allow you to merely refer to them as 'objective 1', or 'objective 2' at later stages in your document without repeating them in full.

The key objectives are those that will be met by the data collection exercise and analysis in the middle of your document. How many objectives? Perhaps just one will suffice. This could be just one variable measured really well. Perhaps two or three objectives are 'do-able' in a dissertation. If there are two or three, they should be linked. If the context of the study is to seek out possible relationships between variables, it is possible to execute three objectives by measuring only four variables. If there were a study commissioned by an industry that suffers from high employee turnover; this would be the DV. Some preliminary work suggests three possible causes (labelled IV1, IV2 and IV3). The four variables to measure are IV1, IV2, IV3 and DV1. The objectives may be expressed thus:

(a)   To determine the impact of distance of travel (IV1) on employee turnover rates (DV1).
(b)   To determine the impact of flexible working hours (IV2) on employee turnover rates (DV1).
(c)   To determine the impact of home working (IV3) on employee turnover rates (DV1).

Alternatively, a study may be around just one IV and three DVs. A ready-mix concrete company proposes to expand its use of water-reducing admixtures to improve workability. It needs to be assured there are no detrimental effects on its products. It commissions some laboratory based experiments:

(1)   To evaluate the impact of varying levels of admixtures (IV1) on long-term durability (DV1).
(2)   To evaluate the impact of varying levels of admixtures (IV1) on compressive strength (DV2).
(3)   To evaluate the impact of varying levels of admixtures (IV1) on cost (DV3).

One reason for your literature review may be to bring your and your readers' knowledge and understanding up-to-date. Should that be expressed as an objective? It is arguably implicit that that is what your document will do and there is no need to express it as an objective. However, you may want to make sure that you express with clarity specific areas of the literature that you will explore, and that is best done by defining those areas in objectives. You cannot be criticised for the latter. If you do adopt this approach, be sure that you make it clear which objectives are being met by the literature review, and which by the main data collection and analysis. In terms of numbering, it may be logical that the literature objectives are numbered first (say 1 and 2), and the objectives to be met in the middle of the dissertation are numbered subsequently (say 3, 4 and 5). To re-iterate, not too many objectives, otherwise the depth of the study may suffer.

There are examples of really good studies which hang together really well, but close to the end of the dissertation process, some hard reflection shows that what has been done in the middle of the document, although excellent, does not reflect with precision the words in the objective. If you find yourself in this situation, on the one hand this is not good, but it is not unusual. If you had been commissioned by a government minister to research and write a report around some given terms of reference, you cannot change the terms of reference if you get to the end of the process and reflect you have done something slightly different;

ministers will not tolerate that. But your dissertation is 'only' a dissertation; it is just your training for research. Whilst ideally you should not be changing your objective towards the latter stages of the dissertation, in discussions with your supervisor you may jointly agree that this is the best thing to do. Having realised you have made a mistake, the strap-line can be altered from 'an objective is a statement of what you will do, and you will "damn well" make sure you do it . . . lest you fail' to 'an objective is a statement of what you have done, . . . lest you fail'.

## 4.5 Objectives with only one variable

An objective may have only one quantitative variable, thus:

(1)   To determine the monetary value range (IV) of new speculative house construction in North West England

The measured outcome of this objective will be quantitative data, which can *(only)* be presented as histograms, bar charts, pie charts, means, medians, modes, etc. (see Chapter 7). An objective may have only one qualitative variable:

(1)   To determine causes of dissatisfaction amongst shuttering joiners

The analysis of objectives with only one variable may be considered insufficient in studies that are required to include a little originality and demonstrate some academic/intellectual rigour. While the qualitative study will not seek to definitely test for relationships, or causes and effects, they may best be directed to teasing out potential relationships.

## 4.6 Objectives with two variables

If data sets are collected about two variables, it may be the case that the measurement of one is relatively easy, e.g. a simple question to ask about gender, or method of work. Alternatively, both variables may need the design of a detailed and comprehensive measurement tool. One data collection exercise may gather information about both variables simultaneously, or it may be that measurement of each variable is undertaken using two completely different methods.

> To compare estate agent opinions about the quality of pre- and post-Second World War houses that are for sale in the North West of England
>
> IV: age of houses, with two values pre-Second World War houses and post-Second World War houses
>
> DV: quality of houses

In this case, the IV may be sourced from secondary data, whilst the DV could be a survey or measurement of opinions.

## 4.7 Hypotheses

Hypothesis (singular) and hypotheses (plural)—there are three common definitions of the hypothesis. Kinnear and Gray (1997) describe a hypothesis as a 'provisional supposition that an IV has a causal effect upon a DV'; and the same authors in 2008 stated 'often a hypothesis states that there is causal relationship between two variables'. Fellows and Liu (2008, p.127) say a hypothesis is 'a statement, a conjecture, a hunch, a speculation, an educated guess . . . which is a reasonable suggestion of a (causal) relationship between the IV and the DV'. Holt (1998, p.16) defines a hypothesis as a 'suggested explanation for a group of facts or phenomena either accepted as a basis for further verification (known as a working hypothesis) or accepted as likely to be true'. The best studies will collect data about an IV and a DV, or collect data about one variable and compare them with data already established as being valid. In the context of studies where the IV and DV are identified, the definition by Kinnear and Gray may be most appropriate.

A hypothesis is only appropriate if the theory has been developed in a significant way from previous work: the hypothesis must not be forced. The hypothesis must be drawn from the theory and the literature. A hypothesis for testing is inappropriate for a qualitative study that seeks to carry out a fundamental investigation to identify what is occurring. A hypothesis may be an output at the conclusion of qualitative work, and it becomes a recommendation for testing subsequently in another study using quantitative techniques.

When writing hypotheses, there are two concepts to be grasped: (a) is the hypothesis written as the null hypothesis or the alternative hypothesis?; (b) is the hypothesis written as a one-tailed or a two-tailed hypothesis?

The null hypothesis is the hypothesis of no association, no difference or no relationship. Dictionary definitions of null are 'not binding', 'non-existent', 'amounting to nothing', 'no elements'. All the n's. The IV does not influence the DV, e.g. interest rates do not influence inflation.

In statistics, tests of hypotheses are executed against the null. Upon execution of a statistical test one of two findings may be expected:

(1)   The IV does not influence the DV, e.g. interest rates do not influence inflation. In this case, the terminology adopted is that the null hypothesis cannot be rejected (this may be interpreted as saying the null hypothesis is accepted but by convention this latter terminology is not used), or;

(2)   Interest rates *do* influence inflation—in this case the null hypothesis is rejected—reject the null—reject the notion that there is no relationship.

The alternative hypothesis is written in a style such that it is suggested that the IV *does* influence the DV, e.g. interest rates *do* influence inflation. A hypothesis intended to be a provisional supposition would seem more appropriately written as the alternative hypothesis.

The concept of whether the hypothesis is one-tailed or two-tailed is an issue of direction—direction of movement in the variables. The hypothesis 'interest rates influence inflation' is written as the alternative hypothesis—but it does not predict the direction of the movement. Will an increase in interest rates cause an increase or decrease in inflation? The writing style for a one-tailed hypothesis could be 'an increase in interest rates will cause a decrease in inflation', or 'an increase in interest rates will cause an increase in inflation'. The hypothesis 'interest rates influence inflation' is alternative and two-tailed.

There are thus four permutations for hypotheses: (a) the alternative, one-tailed hypothesis, (b) the alternative, two-tailed hypothesis, (c) the null, one-tailed hypothesis, and (d) the null, two-tailed hypothesis. A question/objective arises from the articulation of a problem around potential differences in knowledge between younger and older construction professionals about sustainability issues. Possible semantics are illustrated, thus:

- Null hypothesis, two-tailed: no differences exist between younger and older construction professionals in their knowledge levels about sustainability issues
- Null hypothesis, one-tailed: younger construction professionals do not have greater knowledge levels about sustainability issues than older construction professionals
- Alternative hypothesis, two-tailed: differences exist between younger and older construction professionals in their knowledge levels about sustainability issues
- Alternative hypothesis, one-tailed: younger construction professionals have greater knowledge levels than older construction professionals about sustainability issues (or this could be written with older professionals having greater knowledge levels)

Do not complicate the situation. Only one hypothesis is necessary, but you must be clear in your own mind when you write your hypothesis which one of the above it is. It can seem quite odd to commence a study with a statement of the hypothesis written in the null. You may be excited to assert and prove your alternative hypothesis as being definitively correct. The issue of writing a hypothesis as the null or the alternative is only a matter of semantics. It is only words, and you should not worry about it. Therefore, 'interest rates influence inflation' is better than 'interest rates do not influence inflation'. It can be argued that the writing style of the alternative hypothesis is more apt to the 'provisional supposition' definition of a hypothesis. You may wish write your hypothesis as an alternative hypothesis, and put a statement in your dissertation something like 'all hypotheses are written as alternative hypotheses, although it is recognised that statistical testing is undertaken against the null'. However, when you are about to conduct tests in the middle of your document, it may be appropriate to revert to a null hypothesis.

The difference between the two-tailed and one-tailed hypothesis is more important. The rule is that the two-tailed hypothesis should always be used unless there is strong established evidence to predict the direction of movement of variables (Hays 1988, pp.276–277). In the case of interest rates and inflation, such evidence probably does exist and it could be argued that the one-tailed hypothesis is appropriate; a rise in interest rates will lead to a fall in inflation. In many research projects, however, evidence is not available, and the hypothesis tested is often the two-tailed hypothesis. Once your research has begun, you may try and anticipate, if a relationship is found, what direction it will follow. You should still resist the temptation to write one-tailed unless the evidence is already there. Statistically, there is twice the chance of rejecting a two-tailed hypothesis than rejecting a one-tailed hypothesis; that is because your result can fall in the tails either side of a normal distribution (see chapter 7). In some cases, you may find it difficult to predict which way any relationship may go. Take the objective:

to determine whether method of paying craftspeople (IV) affects the timely completion of projects (DV).

The IV has two values: incentive schemes/piecework and fixed wages. On the one-hand, incentive schemes may induce craftspeople to work quickly, hence projects are completed quickly. On the other hand, incentive schemes may induce craftspeople to leave defects behind them, and thus longer snagging periods mean projects are delayed. Which way is it? Do the research and find out.

The wording in hypotheses should imitate the wording in the objectives, which should imitate the wording in the research questions. Students often articulate their research questions, objectives and hypotheses using different terminology. Different terminology in three research questions, three objectives and three hypotheses leave the student attempting nine things; clearly too much. Take a scenario where the problem is based upon female construction professionals who may be demotivated because they feel they are treated less fairly than men. Start with an aim and an articulation of the problem. Then match the research questions *to the* objectives and *to the* hypotheses. The matching process is illustrated thus:

- Working title: is a macho culture influencing the job satisfaction of women working in construction?
- Aim: to establish how to achieve gender equality in the construction industry
- Question; does a macho culture influence the job satisfaction of women working in construction?
- Objective; to determine whether a macho culture is influencing the job satisfaction of women working in construction
- Research hypothesis (alternative): a macho culture is influencing the job satisfaction of women working in construction

If your first study objective is being met by the literature review, a hypothesis may not be appropriate. Therefore give the question and the objective, and clearly state there is no hypothesis for objective number 1.

## 4.8 Independent and dependent variables

A variable is any facet that can have more than one value, or that can exist in more than one form; it is capable of moving. In mathematics, a concept that is not a variable is a constant. Variables may exist in quantitative or qualitative studies. In research we are interested in measuring variables; a useful strap-line may be, 'if it moves, measure it'. Businesses are dealing with variables without labelling them as such; they do not usually use this type of language. In your dissertation you will also be dealing with variables. You should clearly use the language of variables; label, define and measure them.

Another perspective on the concept of variables is that businesses are always looking to change; change is an often used mantra. The change is for a purpose, not just for change's sake. It is to improve something else; that is, if we change 'A', that will impact or influence 'B'. At a personal level we are looking to change things, for example change the way we work to save time expended doing tasks. Whether at a business or personal level, whilst we are implementing change we are 'manipulating' variables over which we have control. We anticipate that by manipulating the variable that we can control (the IV; say, our method of working), this will influence time expended doing tasks (the DV). Hopefully, as we change our method of working, this will be for the better, and we will save time; but we cannot be sure of this until we have done the research, mindful that the change, when implemented, may not work well and things may actually take longer.

Companies strive to optimise performance. The three most often quoted performance measures are cost, time and quality, sometimes called the 'iron triangle objectives'. Added to the iron triangle should be objectives around performance in safety and sustainability. Research projects may seek to tease out what variables are impacting on these business objectives. More specifically, research projects may wish to tease out whether or not one particular variable (the IV) influences one performance measure (the DV). The IV is said to be changed, moved, altered or manipulated, and this will have an influence on the DV; movement of the IV causes movement in the DV. Some researchers use the language of causes and effects. Causes and effects are IVs and DVs under a different name.

Research is not just focused around the above five performance measures. Constructing Excellence in the Built Environment has designed dozens of measures across 13 groups. For example, in the group 'Construction—Consultants' there are eight key performance indicators (KPIs), as identified by Constructing Excellence (2009), thus:

(1)  Client Satisfaction Health and Safety Awareness.
(2)  Client Satisfaction Overall Performance.
(3)  Client Satisfaction Quality of Service.
(4)  Client Satisfaction Timely Delivery.
(5)  Client Satisfaction Value for Money.
(6)  Productivity.
(7)  Profitability.
(8)  Training.

There are many other things research projects may wish to measure, for example, 'soft' concepts used in human resources management, such as employee satisfaction. Research may be commissioned to determine the effect of leadership style (the IV) on employee satisfaction (the DV).

It is perfectly acceptable to have just one variable in your study. It would not carry the label of IV or DV, but merely 'variable'. The study objective may be merely to 'find something out', for example, if it is not already available, to collect data about the gender of people who attend sporting events. It is important that this data set is available to the designers of stadia. However, it can be argued that the better studies contain at least one IV and one DV. An IV and a DV are better in studies because, after all, life and business are often about answering the 'what if' question. What will happen to 'this' if we change 'that'. If 'A' is manipulated or moved, what will happen to 'B'? Having two variables is also important because it gives you the opportunity to carry out more robust analysis, particularly if you are doing a quantitative study. If you have only one variable, your analysis may be merely descriptive; if you have two variables, you can do inferential analysis. Inferential analysis is a little more complex than descriptive analysis. If you can do it well, and demonstrate you understand the principles underpinning some of the simple statistical tests, you have the opportunity to score more marks.

A study to examine the relationship between interest rates and inflation has two variables. A relationship is well established in the UK economy, and thus interest rates are often moved/manipulated on the basis that over a relatively short timescale they influence inflation. The scenario historically has been that if interest rates are moved upwards, inflation will move downwards. In this context, interest rates are the IV and inflation the DV. Since the relationship between these two variables is well established in the UK economy, a more relevant study for construction students may be between interest rates and some measure of construction industry inflation such as indices published by the Building Cost Information Service.

If a relationship is found between two variables, it does not necessarily mean 'cause'. It may be stated that there appears to be an association between two variables, and that association warrants further investigation. There may appear to be an association between unhealthy lifestyle and ill-health in people, but the first is not necessarily the cause of the second. To assert that an IV has a causal effect on a DV may need more work, including some careful preliminary speculation around the 'why' issues'

Sometimes students have two variables in their study, but have difficulty in identifying which is the IV and which the DV. The golden rule is to establish a time gap between the variables; in a timeframe the IV must come first. The IV is manipulated, since we have control over it (e.g. as an economy we control interest rates), and at some later point in time, be it short or long term, this change will impact on the DV. It may be helpful to present the relationship as a diagram in the introduction chapter (e.g. Figure 4.1), and on each occasion, the two variables are mentioned together as in an objective; the IV is written first followed by the DV. For example, if the study were to determine whether the environmental performance of companies (the IV) influences profitability (the DV), it is clearer to write an objective as:

> To determine whether the environmental performance of companies (the IV) influences profitability (the DV)

Perfectly acceptable, but perhaps not so clear is:

> To determine whether profitability (the DV) is influenced by environmental performance of companies (the IV)

The issue is complicated a little by the notion that IVs and DVs can 'melt' or 'merge' into one another; that is, they are acting on each other simultaneously. It could be argued in another study that inflation is the IV impacting on interest rates as a DV. This is because inflation is rising in the early part of an economic cycle, then a decision is made to increase interest rates in the middle of the cycle with the intention of hopefully reducing inflation at the end of

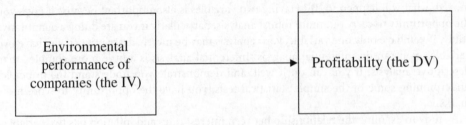

**Figure 4.1** Relationship between the variables in the study objective.

the cycle. The scenario then becomes interest rates influencing inflation, influencing interest rates, influencing inflation and so on.

So whilst the notion of variables complicates things a little, you should try to simplify it in your study. You need to understand that, for some studies, variables 'melt' into each other. Use some words in your dissertation to show that you understand this. But for clarity it is useful to label your variables as IV or DV. Which variable is the IV and which the DV depends on the context of your study as defined by your description of the problem and your objectives. Being clear about which is which helps to focus the whole structure of the research, but do not get 'hung-up' about it. For your study know your IV, know your DV and recognise that in another study they could be classed the other way round. It may be more accurate to say that there is a relationship or consistent association between the variables. A movement in one variable may result in the movement of the other variable; something is going on—there is a link between the variables that needs an explanation. Establishing a link between variables may be used in theory-building exercises.

An academic exercise which examines industrial problems cannot actually solve them. It can, however, identify causes or identify what the variables are, particularly what the IVs are. Developing theories about 'how' and 'why' things happen are assisted by establishing causes. Conclusions may indicate a need to change or manipulate IVs, and recommendations may suggest how this manipulation may take place.

## 4.9 Lots of variables at large; intervening variables

In life, there are lots of variables at large. In social sciences particularly, there will often be many IVs impacting on one DV. A research project may wish to seek out as many IVs as possible, and determine the strength of their influence upon the DV. Some IVs will have a strong influence, others less strong. Some of the more sophisticated statistical techniques are able to test for the influence of more than one IV upon one DV. It is adequate at undergraduate level, in any one test, to look for the effect of one IV on one DV. If a relationship is found, it is not to draw conclusions that are too strong, given that the strength of a relationship may only be modest, and there may be other more important variables that warrant attention. A modest relationship may indicate that significant manipulation of the IV is required to instigate less significant movement in the DV.

In some cases it may be that the terms 'cause and effect' are too strong. Managers have control of IVs that they can manipulate in the hope of changing a DV related to business performance. But it can not always be asserted that a particular IV causes less than optimum performance in a DV. It may be found that a particular diet is associated with few occurrences of a particular illness, but that does not mean that failure to follow that diet is a cause of that illness. It is merely that there appears to be an association between the diet and the illness prevention. Therefore try to use the connecting words between the IV and the DV carefully. You may consider association, influence, impact, relationship, correlation, link, effect, cause.

Variables may be linked by an 'open or closed chain'. A study may initially want to determine whether there is a relationship between variables A and Z, but discovers that there are lots of intermediate relationships or intervening variables. For example, a research project may be initiated to examine the relationship between leadership style and a lead business objective around productivity. After examining the literature, it is apparent that another variable, employee satisfaction, sits somewhere in this equation. Therefore, data sets are collected about three variables, labelled initially:

- Var 1: leadership style
- Var 2: employee satisfaction
- Var 3: productivity

Analysis may show that relationships between Var 1 and Var 2 and between Var 2 and Var 3 are stronger than the one between Var 1 and Var 3. This may be because there are more variables than just Var 1 impacting on Var 2. An outcome of the study could be a recommendation to seek out what other variables are impacting on Var 2 to thus improve Var 3. Could there be other variables between Var 2 and Var 3? This could be another study.

Whether to label Vars 1 to 3 as IVs or DVs can be considered, but should not be made into a big issue. In a test between Var 1: leadership style and Var 2: employee satisfaction, Var 1 may be labelled the IV and Var 2 the DV. But in a test between Var 2: employee satisfaction and Var 3: productivity, Var 2 becomes the IV and Var 3 the DV. Not to complicate things, but in a test between Var 1 and Var 3 there is a temptation to label Var 1 as IV and Var 3 as DV. But could it be the case that it is the other way round; that is, if productivity levels are high, this takes pressure off managers and they feel they can afford to be people-orientated managers. If productivity levels are low they adapt their style to become task-orientated. Is the link between the variables circular, or as a closed chain? Figure 4.2 illustrates potential relationships.

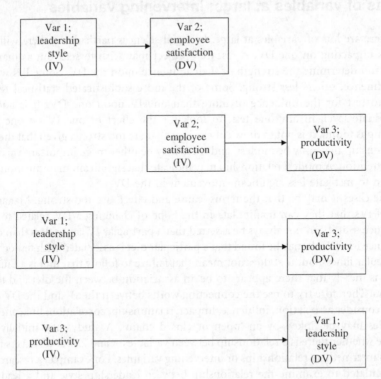

**Figure 4.2** The relationship between variables. IV, independent variable. DV, dependent variable.

Consider another example in construction. The objective for a study could be to determine whether the environmental performance of companies (the IV) is linked to profitability (the DV). Remember there are lots of variables that impact on profitability; you are just trying to tease out whether one of them is environmental performance. It may be hoped that those companies who perform best generate most profits, perhaps through cost savings and/or increased turnover. However, it may be the case that it is the other way round. Companies who invest money in their environmental performance may only increase cost without benefits, and therefore those with best performance make least profits. If the study finds that environmental performance does have some influence (even if only modest) on profitability (the DV), it may follow that those companies with best profitability have confidence and money (profit now as an IV) to invest in improving environmental performance (a DV), which generates more profit which generates more investment in environmental performance, etc.

## 4.10 Subject variables

In your findings you may wish to assert that an IV influences a DV. However, it may be the case that there are other variables impacting on your DV. You cannot measure all the variables that are at large; to measure one variable can be a whole study. You should not attempt two studies in one. If another variable is very important in helping to understand a problem, let that be a separate study. But there will often be some further data that can be collected without putting unreasonable demands on your time or the time of others. If you are collecting data from archives, there will often be a whole host of information available. In surveys you can ask general questions to get data about other variables. These variables may be called 'subject variables'. Typical general questions are about respondent personal data (sometimes called demographic data), e.g. age, gender, job title, ethnic background, qualifications. They are also called 'subject variables'. Possible responses should align with those in authoritative documents, e.g. the UK census. You may also ask about the sector of the construction industry in which respondents work, or the size of company, with possible responses aligned to the construction statistics annual (Office for National Statistics 2009). Two examples are used to illustrate the concept of subject variables.

### *Example 1*

A study may be commissioned to investigate consumer belief in man-made climate change (IV) on their willingness to invest in photovoltaic roof panels (DV1). A questionnaire survey is administered. The researcher also collects data about the subject variable 'whether respondents have children', (SV 1); other SVs could be collected too, e.g. age of respondent, employment status, disposable income, gender, current home type, and whether a home-owner. SV1 has two values; no children, or yes children.

A basket of questions is used to tease out and measure both the IV and the DV. Table 4.1 illustrates a hypothetical data set; for ease of illustration the IV and DV are measured on a scale of 0 to 10, with 10 indicting high belief in man-made climate change and high willingness to

**Table 4.1** A hypothetical data set.

| Area | IV: consumer belief in man-made climate change | DV: willingness to invest in photovoltaic roof panels | SV: whether respondents have children; 1 = no children, 2 = yes children | DV scores repeated for 1 = no children | DV scores repeated for 2 = yes children |
|---|---|---|---|---|---|
| 1 | 5 | 4 | 1 | 4 | |
| 2 | 6 | 3 | 1 | 3 | |
| 3 | 9 | 7 | 2 | | 7 |
| 4 | 8 | 8 | 2 | | 8 |
| 5 | 9 | 10 | 2 | | 10 |
| 6 | 3 | 4 | 1 | 4 | |
| 7 | 9 | 9 | 2 | | 9 |
| 8 | 3 | 5 | 1 | 5 | |
| 9 | 8 | 8 | 2 | | 8 |
| 10 | 4 | 5 | 1 | 5 | |
| | 64 | 63 | | 21 | 42 |
| Means | 6.4 | 6.3 | | 4.2 | 8.4 |

invest. The scatter diagram is shown in Figure 4.3. Figure 4.4 illustrates IVs and SVs impacting on the DV.

Just eyeballing the scatter diagram indicates there may be a relationship between these two variables. Those respondents who have high belief in man-made climate change have high willingness to invest in photovoltaic roof panels. Those respondents who have low belief in man-made climate change have low willingness to invest.

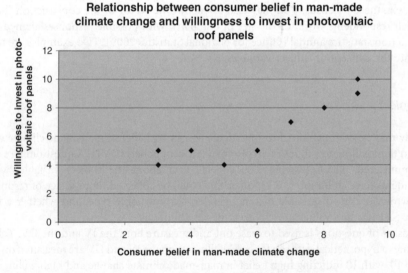

**Figure 4.3** The relationship between the IV and DV, which by eyeball is judged to be a good relationship.

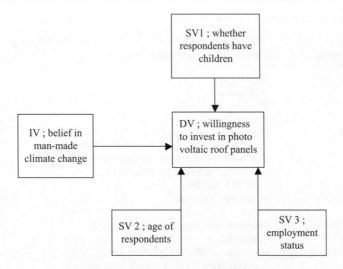

**Figure 4.4** IVs and SVs impacting on the DV.

However, observe the mean scores for willingness to invest for respondents who have no children, and for respondents who do have children. From Table 4.1, those scores are 4 .2 and 8.4 respectively. There is something else going on in this data set. Is it the case that, it is whether consumers have children and not belief in man-made climate change, that influences willingness to invest, the DV?

If this is true, it would be difficult to draw strong conclusions about the relationship between the IV and the DV. The eyeball relationship that is found in Figure 4.3 between the IV and DV could be spurious. As part of your post-hoc analysis, you may assert that consumers who have children are more willing to invest in photovoltaic roof panels than consumers who do not have children.

An alternative data set could show that mean scores for willingness to invest could have been similar for respondents with or without children; say one at 6.4 and the other at 6.2. That lends weight to an argument that the relationship found between the IV and the DV is indeed genuine; it is not spurious.

### Example 2

A study objective is to determine whether leadership style (IV) influences job satisfaction (DV). A questionnaire is used to measure the DV (the IV could be measured by questionnaire too, or alternatively by other means). There is a response from 50 respondents; job satisfaction is measured on a scale of 0 to 10, with 10 being high. The mean score is calculated at 7.50. Analysis to see if results for the DV vary between categories may look like the data set in Table 4.2.

Simple 'eye-ball' analysis shows that for all but one category, the mean scores are similar; that is, these SVs do not seem to impact on the scores for the DV. For the category 'sector', there is a larger difference in mean scores. There is a danger that the relationship found between the IV and the DV is spurious.

**Table 4.2** Mean satisfaction scores measured between categories for typical subject variables.

| Subject variables | Categories in each subject variable | Mean job satisfaction scores |
|---|---|---|
| Age | 16–44 years | 7.40 |
| | ≥45 years | 7.60 |
| Gender | Male, $n = 45$ | 7.45 |
| | Female, $n = 5$ | 7.53 |
| Qualifications | Craft qualification | 7.42 |
| | 2 + A levels, NVQ level 4–5, HNC, HND | 7.51 |
| | First degree or equivalent or professional qualifications (e.g. MCIOB, MRICS, MCIAT, MRIBA) | 7.60 |
| Ethnic background | White British or Irish | 7.41 |
| | Other | 7.55 |
| Job occupation group | Manager, senior official, or professional occupation | 7.39 |
| | Associate professional or technical occupations | 7.61 |
| Sector | Non-residential building | 6.90 |
| | House building | 8.05 |
| Size of company | 1 to 79 employees | 7.49 |
| | 80 or over employees | 7.51 |

If you find a relationship between a SV and a DV, do not throw away the finding about the relationship between the IV and the DV; you may wish to express some caveats in your findings or it may be possible to do some more exploration and analysis of the data. If you go into your research with the 'hope' of finding a relationship between the IV and the DV (though you should not necessarily be doing this), your eureka moment will be when you get significant results; that is, a relationship between the IV and the DV. Another eureka moment should be if you do eyeball analysis around subject variables, and this shows only small differences in means; that is, there is no relationship between the SVs and the DV. This would give you greater confidence that any relationship between the IV and the DV is genuine; it is not spurious. It also indicates that the data set you have is homogeneous; the matrix of numbers that you have hangs together well around an assertion that the IV influences the DV.

Collecting data about subject variables and making observations helps to improve the validity of your study. You could be asked, perhaps in a viva, 'how do you know that it is the IV impacting on the DV, and not some other variable'. Your answer is clear, 'I have collected data about as many other variables as I reasonably could, and (if this is the case) I have found no other relationships'. If you do find relationships between SVs and the DV, do not be disappointed. You will get marks for doing the analysis, but you should also write in your conclusion chapter about how this is a limitation on the validity of your work. If you realise and understand limitations, you can get marks for this too, but if your work is limited and you do not understand this, then marks could be deducted. You may find it appropriate to compare scores of the SV against the IV as well as the DV, especially if it is the case that the issue over which is the IV and which is the DV is open to interpretation. The statistical tests in the above examples to determine whether relationships are significant are the Spearman's correlation (for testing between the IV and the DV) and the Mann–Whitney test (for testing between the SV and the DV). Chapter 8 will illustrate how these tests and probability can be used to determine whether relationships are significant.

## 4.11  No relationship between the IV and the DV

There may be a temptation to hope that, at the end of a study, a relationship is found between an IV and a DV; you may be disappointed if at the end of your journey you find no relationship. Do not be disappointed; whether you find a relationship or not is a factor in judging the success of your study. The idea is not to force relationships, so that after analysis you can assert 'I told you so'. The purpose of research is to determine whether there is a relationship between variables and to tease them out gently. If relationships are found, they should be re-tested using different methodologies to assure validity. Society needs to understand these relationships to build its knowledge. We need to know whether your IV influences the DV; or is it another IV? Society needs to establish the links from variable A to variable Z, and it can only make those links in very small, often minute steps over long time periods. It cannot be that every piece of research finds relationships; that would be preposterous. The process is like going through a maze. We want to get from start to the finish, and that journey involves linking a chain and a web of variables. It is inevitable that research will take the wrong route in the maze; it may explore possible relationships and find none. The wrong road you have taken must be marked, you must turn back and try a different route. The wrong road is marked by publishing your work, and telling other researchers that you have tried this route and it is a dead-end. Following researchers can then devote their attentions to other methodologies or other variables. Studies that do find relationships between variables may have conclusions written with vigour, championing the study as though a victory was won. If you do not find a relationship, you must not let your conclusion suffer. You should still write it with vigour, perhaps with recommendations to suggest slightly different methodologies for re-testing, or suggesting slightly different variables. It is often the case that the studies that grab media headlines are those that find relationships between variables. So be it; you are not after media headlines from your dissertation work.

## 4.12  Designing your own measurement scales

Your dissertation is about measuring something really well. You should not try to measure too many things. To measure a few things well should attract good marks. To measure many or complex things superficially will not attract good marks. For example, it would not be possible to measure in one dissertation the success of companies around the cost, time and quality triangle. Better to just measure one, and indeed just one small part of one of these.

Qualitative analysts may argue that they are not about measuring things. Their approach is more about gaining insights into the world and about how other people see things. Qualitative analysts collect rich data. Certainly an outcome of qualitative research is not to score things on scales. If research is about the study of constants, that is correct in an absolute sense. More likely that most qualitative research is about things in life that are changing and are therefore variables. It is still not about scoring things on scales, but it is about collecting evidence and making judgements one way or the other about issues. An outcome of research may be that something is allocated the qualitative label of 'good' or 'bad'. This outcome could have gone one or two ways; it is variable.

Some things that you measure do not need new measurement designs, nor may it be appropriate to adapt them to 0–10 or 0–100 scales. Some examples are:

- Age of construction workers 16–65 +
- Interest rates; normally 0 to say 10%—possibility of negative rates or rates well over 100% in unstable economies
- Profit on turnover; normally −25% to + 25% but unlimited extremities possible (note: lots of other financial and accounting measures/ratios)
- Compressive strength of concrete (also workability, water/cement ratio, cement content per m³.)
- Tensile strength
- Density
- Number of people employed in UK construction
- Heights of the world's tallest buildings
- Bulking by volume of excavated stiff clay
- Temperature
- Thermal conductivity
- Illumination levels
- Noise volumes
- Moisture content
- Absorption
- Percentage aggregate passing a sieve

In many spheres of research it is not for you to design new things and new ways. It is far better to use established authoritative methods to help you. In the context of measuring your variables, delve deeply into the literature to see if a measurement tool has been used before. If necessary, adapt authoritative material to the context of your study. Using the work of others to provide a platform for your work improves its validity, and should give you more marks. Where you do use or adapt the work of others, make sure you cite the original source. If you adapt the work of others, cite the material you use as 'adapted from Smith (2010)'.

There are, for example, lots of questionnaires in the literature about leadership style. To design your own would be foolhardy. On a continuum, some leadership scales are anchored at the extremities by terms such as 'task orientation' and 'people orientation'. Managers may be judged to be at one end or the other, or somewhere in between. Through web-based searches there are lots of questionnaires available, using key word phrases such as 'leadership style questionnaire'. Do not just use any questionnaire that you stumble across. Browse them all and pick the most authoritative source that is relevant to the context of your study, adapting it if necessary. Better than just using the web, also look at what may have been used in some of the academic journals. Make sure you do not breach any copyright.

Many modern KPIs are designed around zero to 10 (0–10) scales. If you do design your own scale, it is logical to do the same or possibly expand it to the range 0–100 like an examination scale. Note that the scale should be anchored at zero and not one. A student who fails to write anything in an examination will get 0% and not 1%.

Some things to be measured may not lend themselves in their origin to the decimal scale, but can be converted to it by simple arithmetic. For example, a questionnaire may be used to measure the commitment of speculative housing bricklayers to producing high quality work. There may be eight statements, to which bricklayers may tick one of six possible responses, e.g.:

- Q1. I make sure that my tools have been cleaned after use
- Q2. I am careful to leave enough time to point or joint my work
- Q3. I only lay to heights that I can comfortably reach
- Q4. I ensure that my work is covered to protect it from weather if necessary
- Q5. I ensure I am working from a firm level platform
- Q6. I am especially careful when setting out, building corners and in plumbing work
- Q7. I ensure that cavities are clear of mortar droppings
- Q8. I am careful to ensure that ties and damp proof courses are installed correctly

The possible responses are: always, usually, sometimes, rarely, never. Ensuring that a 'never' response scores zero, these responses would be given numbers thus: never $= 0$, rarely $= 1$, sometimes $= 2$, usually $= 3$, always $= 4$.

For the eight questions the possible range of scores would be from a minimum of 0 (eight questions multiplied by a lowest score of 0 for all questions) to a maximum of 32 (eight questions multiplied by a top score of 4 for all questions). If the scores for an individual bricklayer were 23 out of 32, on the face of it that might seem to be an 'ok' score. But to simplify it, do some arithmetic to convert it to a 100-wide decimal scale, thus:

$23/32 \times 100 = 71.87$, say 72%

To emphasise the concept of using a zero anchor, suppose your possible responses were scored as follows: never $= 1$, rarely $= 2$, sometimes $= 3$, usually $= 4$, always $= 5$. For the eight questions the possible range of scores would be from a minimum of 8 (eight questions multiplied by a lowest score of one for all questions) to a maximum of 40 (eight questions multiplied by a top score of 5 for all questions). Imagine the possibility of bricklayers who have no concern for quality, perhaps only concerned to maximise earnings. Those bricklayers would score 8 out of 40. Convert this to the decimal scale ($8/40 \times 100$) and this becomes 20%. These hypothetical bricklayers logically should not score 20%, they should score 0%. Therefore, for the above a scale of 0–4 is better than a scale of 1–5. A hypothetical set of results are shown in Table 4.3, with only 10 respondents scores showed for brevity. Many descriptive statistics are possible on a data set such as this (see Chapter 7); again for brevity, just the mean is shown. The overall commitment to quality of 60% is less than desirable. Question 1 scores highly, Q5 and Q7 score poorly. These individual scores could be worthy of discussion. Bricklayer 10 scores full marks, and in practice that is hopefully not unusual. Bricklayer 9 may have hopefully given rogue answers, and they should arguably be removed from the analysis. You may reflect on the questionnaire in the discussion; should Q6 really have been three separate questions and Q8 two questions?

There is an issue about whether to give the 'best' response a high or low number. Following the principles of construction KPIs 0–10 scale, the best score $=$ high number. This seems logical; having a scale the other way round, for instance with the worst bricklayers

**Table 4.3** An example of a self-designed measurement scale; commitment of speculative housing bricklayers to producing high quality work.

|  | Q1 | Q2 | Q3 | Q4 | Q5 | Q6 | Q7 | Q8 | Q1 to Q8 total | Percentage score |
|---|---|---|---|---|---|---|---|---|---|---|
| R1 | 4 | 4 | 2 | 3 | 2 | 3 | 2 | 3 | 23 | 72 |
| R2 | 4 | 3 | 3 | 2 | 2 | 3 | 1 | 2 | 20 | 63 |
| R3 | 4 | 3 | 2 | 3 | 1 | 2 | 1 | 4 | 20 | 63 |
| R4 | 3 | 2 | 2 | 3 | 1 | 1 | 1 | 2 | 15 | 47 |
| R5 | 4 | 3 | 4 | 3 | 1 | 3 | 0 | 2 | 20 | 63 |
| R6 | 4 | 2 | 3 | 3 | 2 | 3 | 0 | 3 | 20 | 63 |
| R7 | 4 | 4 | 3 | 3 | 2 | 3 | 1 | 3 | 23 | 72 |
| R8 | 3 | 2 | 2 | 2 | 2 | 2 | 2 | 2 | 17 | 53 |
| R9 | 0 | 0 | 0 | 0 | 0 | 0 | 0 | 0 | 0 | 0 |
| R10 | 4 | 4 | 4 | 4 | 4 | 4 | 4 | 4 | 32 | 100 |
| Totals | 34 | 27 | 25 | 26 | 17 | 24 | 12 | 25 |  | 596 |
| Mean | 3.4 | 2.7 | 2.5 | 2.6 | 1.7 | 2.4 | 1.2 | 2.5 |  | 59.6 say 60% |

scoring 100% instead of 0%, seems illogical. Therefore, in the above example, do not score always as 0 and never as 4. However, be mindful that in some audits, the best score is a low number. Government audits of public bodies have gone both ways, e.g. in one sector the scale may be: 1 = excellent, 2 = good, 3 = adequate and 4 = inadequate, but in other sectors the analogy is to stars (as used to classify hotels), with five stars being excellent and possibly expanded to six stars as outstanding. It is not always the case that, on a given spectrum, one end is necessarily better than the other, for example again, leadership styles. It is irrespective whether task orientation is scored at 10, and people orientation at 0, or the other way round.

If you are doing some kind of survey, when you give respondents a choice of responses to questions or statements, you must consider whether to allocate a number or a word. Particularly if it is a narrow scale, e.g. five wide rather than 10 wide, it is better, if possible, to allocate words or descriptors. Rather than asking bricklayers about how often they clean their tools on a range of 0 to 4 with 0 being never and 4 being always, it is better to give only words (always, usually, sometimes, rarely, never), and then you as a researcher allocate the numbers later.

Some thought needs to be given to how scales are anchored. Using a sporting analogy, imagine a project to measure the ability of footballers. This may be conducted by a football scout who watches games in the local community. The scout scores the footballers on a scale of 0 to 10. Ten is anchored with lots of descriptors, summarised by 'the potential to play in higher leagues'. Footballers who may score 10 out of 10 playing at a local level are clearly excellent players. But now take a project to measure the best footballer in the world. On this scale our local community player will likely score only 1 or 2 out of 10, since the descriptors for 10 out of 10 will be anchored by words such as 'world class'. In a construction context, if you are measuring the performance of companies around health and safety, is the top score reserved for best UK performance or is 10 out of 10 only for world class performance? Is UK performance in health and safety world class; are they the same thing? There is an argument that smaller companies find it difficult to perform at the same levels as larger companies in health and safety. If that is correct, and your study is around the health and safety performance of smaller companies, perhaps 10 out of 10 is reserved for best UK performance of small companies. If you undertake a study of safety performance in economies that are less well developed than the UK, it may or may not be appropriate to anchor 10 out of 10 with a descriptor such as 'world class'.

Some designed authoritative scales are not on the range of 0–100. For example, one measure of safety by Constructing Excellence is number of accidents per 100 000 man hours worked. Other measures of safety are around incidents (including near misses) not just accidents.

## 4.13 Levels of measurement

On an everyday basis we collect data to measure things. Variables can be measured at different levels, thus: (a) categorical (nominal), (b) ordinal, (c) interval, and (d) ratio. The sequence is important: the richness of the data and the power of statistical tests that you may do increases as the scale moves from (a) to (d). Categorical variables are also called 'nominal variables' in many texts; you need to be able to use the two descriptors interchangeably. The sequence of the variables may be remembered by the mnemonic COIR (the fibre from the outer husk of the coconut) or if you prefer to use the term nominal, NOIR (in French, NOIR is the translation of the colour black when used with masculine nouns).

To illustrate with an example; if student performance is measured, this may simply be a frequency count for the number of students who pass and the number who fail. This data set is at the 'categorical' or 'nominal' level. Imagine that you were told you had passed an assessment; you are in the 'pass' category. Perhaps your initial reaction would be 'excellent', but then after a while you may wonder whether you only just passed or whether you passed comfortably. You want 'richer' information. You therefore go to your assessor and ask for more details about your pass; you are told you have a grade 'B'. This grade is now at the ordinal level; the possible grades are 'A', 'B', 'C', etc. Emphasis is on the word 'order', since 'A', 'B', 'C' are in an order. Your reaction is again 'excellent', but then after a while you may wonder whether you only just obtained a grade 'B' or whether you nearly achieved the higher order grade of 'A'. You therefore go to your assessor and ask for more details (richer information) about your grade 'B'; you are told you have 65%. The latter is at interval level. It is not possible to distinguish the true distance between the ordered grades 'A', 'B', 'C'. Lying behind the ordinal letters may be the interval numbers A = 70%, B = 69%, and C = 50%. In this case, the distance between the A and the B is only 1%, but the distance between the B and the C is 19%. The ordinal grades are masking the richness of the interval numbers. Examining Boards often mask the richness of the data they hold. Student scripts are often allocated a percentage mark at the interval level, but Boards only publish results at the ordinal level. Students might think this unfair. In some instances, collapsing data from the interval to the ordinal, and even categorical level, may be justified, and indeed necessary. Interval data may be too much to digest in a world full of information overload. Whilst government holds information about percentage marks for students at the interval level, the headline data for GCSE results is published at the categorical level; that is, the frequency count for the number of students who gain five passes at grade C or above. There are two categories; either students get the five passes or they do not.

When you collect data, for some variables you may have a choice of whether to collect it at the categorical, ordinal or interval level. Get the data at the richest level possible. Getting the data at the highest possible level can affect the validity of your dissertation. However, there may be good reasons for not seeking information at the highest levels. People may be reluctant to give details at the interval level about age (precise age in years) or salary (precise earnings in £). Also ethically, you may consider it inappropriate to intrude on information that respondents are reasonably entitled to keep private. You may, however, tease out details at the ordinal level (age 25–37 or 37–55, etc., salary £20k to £30k, etc.).

Some students launch into comparative studies as a result of their first practical experience of some facet of their work. This may be a new method of procurement to them, e.g. the private finance initiative, or a new method of build, e.g. timber frame, or a new piece of architectural software. On the one hand comparative studies are fine (e.g. comparing the old software to the new software), but if an underlying 'aim' for your study is for you to become expert in this new facet of your work, perhaps it is better to just concentrate on some sort of appraisal of the new. If you do a comparative study, do not attempt too much. To compare only two facets may be enough, e.g. steel and concrete: do not include timber as a third.

## 4.14 Examples of categorical data in construction

Categorical variables do not have a relationship between each other that can be measured; they are different and there is no distance between them. 'Things' are put into categories or boxes; they are not on a continuum. Apples, oranges and bananas; all are different. Measurement within each category is by frequency counts. Table 4.4 illustrates some variables

**Table 4.4** Examples of categorical data in construction.

| Variable | Category W | Category X | Category Y | Category Z |
|---|---|---|---|---|
| Gender | Male | Female | | |
| Student assessment | Pass | Fail | | |
| Colour of company logo | Blue | Green | Yellow | |
| Direct question; is your company registered to ISO 14001? | Yes | No | | |
| Type of material | Steel | Concrete | Timber | Plastic |
| Size of company | Large | Small | | |
| Method of procurement | Competition | Negotiation | | |
| Sector of construction | New housing | Other new non-housing | Repair and maintenance | |
| Type of employer | Client | Consultant | Contractor | |
| Method of house frame construction | Traditional/ masonry | Timber | Steel | |
| Type of cement | Ordinary Portland | Masonry | | |
| Architectural style | Ancient Greek | Roman | | |
| Occupation | Director/senior manager | Operations manager | Technical or service support | Other |

that may be measured at the categorical level in construction research. Whether they are communicated in the sequence of the allocated labels W, X, Y, Z or say Z, Y, X, W is irrespective. There is no relationship between the categories; one category cannot be considered to be closer to one neighbour than it is to another.

Some of these variables by their very nature can only ever be measured at the categorical level, e.g. gender, and there can only be two categories. However, some of these variables could be measured at ordinal or interval levels; the issue may be a little grey. As materials, steel is steel and concrete is concrete; clearly two separate categories and irrevocably categorical data. But if the variable was 'method of frame construction', and it has two values, (a) steel, and (b) concrete, some buildings contain a mixture of both. Also, if there were a study to compare competitive bidding and negotiation as methods of obtaining work, for the purposes of the study the variable is 'method of obtaining work' and it has two values or two categories: (a) competitive bidding, and (b) negotiation. There is the possibility to open up this variable to the ordinal or even interval level, since some work may be obtained using elements of both competition and negotiation. It might be that opening up the variables 'method of frame construction' or 'method of obtaining work' to measurement at the ordinal or interval level may form the basis of a better study. However, if the basis of your problem and objective best requires measurement at the categorical level, this is what you should do. You should also defend your decision to do this in your methodology chapter.

## 4.15 Examples of ordinal data in construction

Ordinal variables are placed in sequence, but the distance between ordered categories is not measurable. The distance between 'A' and 'B' may not be the same as the distance between 'B' and 'C'. The sequence of A, B, C is important. In statistical analysis, real interval numbers may be subsequently allocated to each ordered category, e.g. 4, 3, 2, 1, 0 allocated to 'always', 'usually', 'sometimes', 'rarely', 'never'. Fink (1995, p.50) states ordinal data are habitually treated as though they were numeric data. However, some purists may argue that allocating real numbers to ordinal is inappropriate. But there is a need to simplify data; it serves to convert some qualitative judgements or statements by respondents to the quantitative. Therefore, perhaps the leap is justified, providing it is acknowledged by analysts.

Some examples in Table 4.5 illustrate how variables measured at the categorical level in Table 4.4 can be opened up to the ordinal level. In some cases it is illogical to ask opinion questions that permit only responses at the categorical level of yes or no. Often issues are not black or white; there is grey in between.

When allocating labels, e.g. outstanding, excellent, or deciding bands between the ordered categories, it is not for you to decide the labels or the positions of bands. You should use authoritative sources. Sustainability ratings of outstanding, excellent, etc. come from BREEAM (Building Research Establishment Energy Assessment Method). Age bands 18–25, etc. are those used in the UK census. Competitive dialogue is a term used by the Olympic Games Delivery Authority. Size of company is specified in the UK Construction Statistic Annual published by the Office for National Statistics.

**Table 4.5** Examples of ordinal data in construction.

| Variable | Category A | Category B | Category C | Category D | Category E | Category F |
|---|---|---|---|---|---|---|
| Student assessment | Grade A | Grade B | Grade C | Grade D | Grade D | Fail |
| Direct question: is your company registered to ISO 14001? | Yes | No, but we have applied or are close to application | No, but we have the medium-term vision to apply | No, but we have the long-term vision to apply | No, and we have no vision to apply | |
| Size of company | 1–13 | 14–79 | 79–200 and over | | | |
| Method of procurement | Competition | Competitive dialogue | Negotiation | | | |
| Sustainability rating | Outstanding | Excellent | Very good | Good | | |
| Method of house frame construction | All traditional/ masonry | Mostly traditional/ masonry; some timber frame | 50:50 traditional/ masonry; timber frame | Mostly timber frame; some traditional/ masonry | All timber frame | |
| Frequency of wearing PPE | Always | Usually | Sometimes | Rarely | Never | |
| Quality of workmanship | Excellent | Very good | Good | Satisfactory | Fair | Poor |
| Occupation | Director | Senior manager | Operations/site manager | Assistant manager | | |
| Age | 18–25 | 25–37 | 37–55 | 55 + | | |

## 4.16 Examples of interval and ratio data in construction

The values for interval level measurements are placed in a sequence, and the differences between the values are assumed to be equal. For most statistical tests, data at the interval level is treated similarly to data at the ratio level. Therefore, it is not appropriate here to dwell too long about the differences between the two. A key principle is that the zero point on the interval scale is arbitrary, but on the ratio scale it has real meaning. This difference is illustrated in the literature by comparison of the Celsius and Kelvin temperature scales. On the Celsius scale, the freezing point of water has an arbitrary value of zero, and it is thus considered interval, whereas the Kelvin temperature scale has a non-arbitrary zero point of absolute zero, which is denoted °K and is equal to $-273.15$ degrees Celsius. The °K point is non-arbitrary, as the particles that comprise matter at this temperature have zero kinetic energy.

The width of ordinal scales, based on subjective judgements, can be widened from say five wide (excellent, very good, good, satisfactory, poor) to say 11 wide, that is 0–10. As the scales become wider, there is a temptation to consider them as interval data. This temptation is more compelling if the measurement of a variable comprises a basket of measures or multiple item scales. The consequence of upgrading the level of data classification from ordinal to interval is that it allows some of the more powerful statistical tests to be used in analysis.

Each point on the 11 wide scale may have a qualitative descriptor allocated to it. Alternatively, criteria may be designed against a range or band of numbers, e.g. qualitative criteria for the ranges 0–3, 4–6, 7–8 and 9–10. When making decisions about which score to allocate, firstly a judgement is made about which band is appropriate, then whether the item being scored is at the lower or upper end of the band.

A consequence of the difference between interval and ratio data is that for interval data, since ratios between numbers on the scale are not meaningful, operations such as multiplication and division cannot be carried out. However, ratios of *differences* can be expressed; for example, one difference can be twice another.

Many variables that are measured in construction, surveying and science are at the ratio level, e.g. money, distance, mass, speed. Some of the examples of ordinal data in Table 4.5 can be opened up to measurement at the interval level, with real numbers, as illustrated in Table 4.6. Table 4.7 illustrates variables that may be measured as part of research projects. Table 4.8 illustrates three variables that could be measured at the categorical, ordinal or interval levels.

**Table 4.6** Examples of interval and ratio data in construction.

| Variable | Measures |
| --- | --- |
| Student assessment | Possible percentage score 0–100 |
| Size of company | Numbers of people employed, turnover in £s |
| Method of procurement | Competition level 0–10 |
| Sustainability rating | Possible percentage score 0–100 |
| Frequency of wearing PPE | Possible percentage score 0–100 |
| Quality of workmanship | Possibly 0–10 or percentage score 0–100 |
| Age | In years |
| Level of client satisfaction KPI | 0–10 score |

**Table 4.7** More examples of variables to measure.

| Level of measurement | Variable | Number of values | Labels for each value |
| --- | --- | --- | --- |
| Categorical | Method of payment to craftspeople | 2 | Incentive scheme<br>Fixed weekly wage |
| | Systems of working | 2 | In source<br>Out source |
| | Method of placing concrete | 2 | In-situ<br>Precast |
| | Method of drawing | 2 | Manual<br>Autocad or similar |
| | Method of energy generation | 2 | Fossil fuels<br>Renewables |
| Ordinal | Design aesthetics | 3 | Extremely attractive<br>Attractive<br>Unattractive |
| | Completion of projects to time | 5 | Well behind time<br>Behind time<br>On time<br>Ahead of time<br>Well ahead of time |
| | Life expectancy of external cladding | 4 | 20–30 years<br>30–40 years<br>40–50 years<br>Over 50 years |
| | Motivation level of maintenance workers | 4 | Very good<br>Satisfactory<br>Unsatisfactory<br>Poor |
| | Accessibility of buildings | 6 | Excellent<br>Very good<br>Good<br>Satisfactory<br>Unsatisfactory<br>Poor |
| Interval/ratio | Cost of buildings | Unlimited | |
| | Tender price indices | Unlimited | |
| | Tensile strength of steel | Unlimited | |
| | Quality of service | 0–10 | |
| | Compliance with best practice | 0–10 | |

## 4.17 Money as a variable

Whatever your construction discipline, money cannot be ignored. It is not expected that money will necessarily be a key strand of your work, but it is arguably remiss of you if you did not give money at least a peripheral mention. Designers may consider the technical or aesthetic merits of methods of construction. Surveyors may appraise methods of maintaining

**Table 4.8** Examples of variables that could be measured at either the categorical level, ordinal level or interval/ratio level.

| Variable | | |
|---|---|---|
| Interest rates | High/low | Categorical |
| | Very high/high/medium low/very low | Ordinal |
| | 15%, 14%, 13%, etc. down to 1% | Interval/ratio |
| Class sizes | Large/small | Categorical |
| | Very large/large/medium/small/very small | Ordinal |
| | 40, 39, 38, etc. down to say less than 10 | Interval/ratio |
| Directors' salaries | High/low | Categorical |
| | Very high/high/medium/low/very low | Ordinal |
| | Unlimited range | Interval/ratio |

facilities. A finding in favour of one method or another should at least lead to a recommendation to investigate money in another study.

It may be that money is a key variable in your work. If this is the case, how is money defined and measured in the context of your study? It could be on the basis of capital costs, running costs, life cycle costs, budgets, profits, cash flow, retention, value, turnover, return on capital employed, liquidity, gearing, or price earnings ratio. Measures could be compared as £s, percentages, ratios, unit rates. You may wish to study currency exchange rates, and their impact on material prices or delivery lead periods. What do monetary measures include and exclude? Perhaps they include site overheads and do not include head office overheads, or perhaps they include neither or both. Costs may or may not include interest charges, discounts, professional fees, cost of land, and value added tax. Profits may be before or after tax or exceptional costs.

Lowest cost to one party in the supply chain may not be the same as cost to another. Consider a specialist plastering contractor 'A' whose labour, plant, material, preliminary and head office costs for a square metre of plaster are £10.00/m$^2$. The specialist adds 10% for profit to give a gross rate of £11.00/m$^2$. Specialist 'B' may have total costs of £10.20/m$^2$ but adds only 5% for profit to give a gross rate of £10.71/m$^2$. Specialist 'A' has the lowest costs, but clients who employ specialist 'A' will have higher costs than those who employ 'B'.

There is a lot of cost information in the public domain through price books and the Building Cost Information Service. If you are to study the financial performance of other companies, you will only be able to get information that is available to the public. The construction and national financial press summarise the performance of the companies in the construction sector. Those companies that are public liability companies (PLCs) are required to publish annual reports detailing financial performance. Companies that are privately owned (private limited liability, limited liability partnerships or partnerships) are required to submit returns to Companies House for taxation purposes. Information for the larger, privately owned companies finds its way easily into the public domain. If you want details about the financial performance of smaller companies, you may have to go to Companies House yourself.

Companies are very unlikely to give you access to figures that arise from internal management accounting systems. You will not get details about profit figures, valuations, or final accounts. However, as a part-time student, perhaps the accounts of your company are open to you. You may be able to get valid data to use for analysis, provided you ensure the source of the data cannot be traced and there is anonymity and confidentiality.

## Summary of this chapter

The study aim is supported by research questions, objectives and hypotheses. Questions, objectives and hypotheses should all imitate each other. It may be preferable to use 'objective' as the key term around which dissertations revolve. Objectives should be clear, and words within them should not change as the dissertation progresses. All parts of the study should hang around the objectives, which may be written as strap-lines. In your introduction chapter you should identify the variables in objectives and write a narrative to define them in the context of your study. If you have two variables in an objective, identify one as the IV and the other as the DV. Whether your study is qualitative or quantitative, you should focus on measuring something. You may gain more marks if you are able to observe the impact of subject variables on your study. On some occasions it is self explicit how variables are measured, e.g. interest rates, but for some variables you may need to design your own measurement scale. Use authoritative scales wherever possible. Quantitative studies should be clear about the level at which the data set is measured, e.g. categorical, ordinal or interval/ratio. You should not be disappointed if you do not find a relationship between your variables. If money is a variable in your study, be mindful that it is not always best measured simply by cost or profit.

# 5 Methodology

The objectives of each section of this chapter are:

5.1 Introduction: to provide context for research design
5.2 Approaches to collecting data: to classify approaches and give examples
5.3 Types of data: to distinguish between primary and secondary data, and between objective and subjective data
5.4 Questionnaires: to provide guidelines for design, with examples of closed ordinal scales and ranking questions
5.5 Other analytical tools: to provide ideas for analysis that do not fall into the traditional qualitative and quantitative schools
5.6 Incorporating reliability and validity: to explain these concepts and describe how they can be achieved
5.7 Analysis, results and findings: to distinguish between them as a platform for writing this chapter in the dissertation

## 5.1 Introduction

Your methodology chapter should thoroughly describe the way you have conducted your study, so that it may be replicated by others; you should also justify the way you have done it. You need to answer the questions 'Why, how and what'. Often students do not write at length in this chapter. It is as though they think it is 'obvious' what they have done; it may be obvious to the student and supervisor since they have 'lived' the dissertation together, but it also has to be clear to a stranger and new reader of the dissertation. Whilst your dissertation is about you becoming expert in your subject area, it is also about you having insight and a generic appreciation of the role of research in society. It is in the methodology chapter that you can demonstrate that you have these attributes. The necessity to justify your methodology means that you must understand several approaches to dismiss those that are not appropriate. This does not mean that you must be able to use and apply all research methodologies, but you do need to have some depth to your knowledge.

*Writing a Built Environment Dissertation: Practical Guidance and Examples.* Peter Farrell.
©2011 Peter Farrell. Published 2011 by Blackwell Publishing Ltd.

Methodology is about why you designed your research at a strategic level. It is about the generic principles underpinning the process you followed and about your choices. The method itself is at an operational level, and describes the mechanics of what you have done. When designing your methodology, you must remain focused on your objectives. Mindful of resource constraints, the methodology must be the best way of meeting objectives. Methodology is not just about the way you collect your data. It is about the whole dissertation from beginning to end, including how you founded the objectives from the problem, how you conducted the literature review, pilot studies, data collection, analytical methods and the process of developing findings and conclusions. There is a lot of potential for weaknesses in the methodology of your research to invalidate it and for it to be fundamentally flawed. Studies are invalidated for example, if populations are not selected correctly, or inappropriate methods of analysis used, or if there is poor design of surveys. You need to detail what steps you have taken to maximise the reliability and validity of your study.

You must describe in detail each step along the way. Your findings will not be 'believable' if you do not describe your method. You may be familiar with describing methods used in scientific experiments in laboratories. The principles of description are similar in social sciences research. There may be potential from research findings for investment and application in industry. As one piece of research is published, someone else may wish to devote substantial resources to developing that work. It may be that research contains mistakes, or that unreasonable assumptions were made. Subsequent researchers may be able to make initial judgements, by reading your method, whether what you have done is valid. If their preliminary judgements are positive, they may wish to replicate the research 'in the field or laboratory', to validate it truly. Retesting, repeating and replication will take place before business people will commit to any investment.

Your justification must be robust. It is not just to describe the way you have done it, and that is that. Why have you done it this way? What alternatives did you consider? The design you used should be based on authoritative methodologies in the literature. When you read about the subject matter of other people's research, you should also digest their methods. You should read specialist methodology textbooks. Cite all your sources. Your methodology chapter may include statements something like 'the method was adapted from the work of Smith (2010, p.32) and Jones (2009). The xyz approach advocated by Baker (2008) was considered but rejected because . . .'.

If you have performed experiments in laboratories, they may be based on British Standards. Cite the standards, and describe if and why you may have deviated from them. Describe the equipment used. If you have observed a production process, describe the process itself and the precise circumstances of your observation.

If you have asked questions in interviews or surveys, define/quantify your population and describe how you picked your sample. What was the size of your sample and what response rate did you get? What alternative sampling methodologies were considered? Give each question and justify why you have asked it. Which objective is each question related to? Have questions been adapted or taken verbatim from another source? If so, this is fine, but cite your source. If interviews were undertaken, describe how they were conducted. How were introductions handled, how long did the interviews take, were they formal/informal, how were data recorded, etc.? If questions were administered other than by face-to face, how was this undertaken? Possibly by hand delivery, post or electronic web surveys.

Whether you have gone down the qualitative or quantitative routes, the data analysis process must be described and you must substantiate why a particular tool was selected. There is a variety of qualitative approaches suggested by the literature; why have you used the approach of one author and not another? If you used a quantitative approach, did you limit yourself to only descriptive analysis? If you were able to undertake inferential analysis, did you use non-parametric or parametric tests?

You should describe what steps you have taken to ensure that you comply with ethical codes, and if appropriate, undertaken risk assessments. In better studies methodology chapters will write at length about what steps have been taken to ensure that the study is as reliable and valid as is reasonably possible, and will demonstrate genuine understanding about these concepts.

## 5.2 Approaches to collecting data

Bell (2005) suggests seven approaches to research: (a) action research, (b) case studies, (c) surveys, (d) experiments, (e) ethnography, (f) grounded theory, and (g) narrative enquiries. The definitions vary between authors, and the boundaries are therefore somewhat blurred. They may be called approaches to collecting data, and are applicable if the data to be collected are qualitative or quantitative. These classifications are not closed boxes. You may collect some data that could easily fit into both categories; you may chose not to place your collection method emphatically into one box. You need to be a realist in whichever approach you use; some data may just be too difficult to get at. But you should also be innovative and proactive; not just follow the usual methods, e.g. surveys, nor be passive. Construction projects hold a wealth of information suitable for analysis such as production data, minutes of meetings, diaries, photographs, contracts, correspondence files, cost reports, drawings, specifications, material receipts, time sheets, time and motion data.

A word of caution when doing quantitative analysis based on data from action research or case studies; it is possible when taking information from one source, perhaps your own company if you are a part-time student, to get large data sets that present themselves as suitable for quantitative data analysis. It is perfectly acceptable that you do the quantitative statistical work, but with the proviso that you do not infer that your findings would be replicated in the wider world; only surveys that use a representative sample from the whole of the population allow you to make inferences.

Action research is a possibility for part-time students in construction, acting as an 'insider'. Alongside your everyday work activity you may simultaneously collect data. For example, you may undertake a qualitative or quantitative study to investigate the leadership style of chairpeople on the effectiveness of meetings. As a participant in these meetings, you may take notes that allow you to action any items that fall on you in your job role, but you can also collect data (possibly tape recorded) that measure the two variables 'leadership style of the chair' and 'effectiveness of the meeting'. Alternatively, you may investigate the impact of amendments to standard forms of building contract on the success of projects, again either qualitatively or quantitatively. Archived files from completed projects could be opened, and each project evaluated against each variable. Interviews could be held with professional staff who worked on the projects to enhance the validity of judgements about success. Whether this example is action research or a case study is irrespective; perhaps it falls somewhere between the two.

Case studies themselves can be undertaken as an insider, or you may be given access to a case by another organisation. There is also the increasing possibility that there may be sufficient data available in the public domain about a past or current case. This gives full-time students the opportunity to do applied work that they may feel is mostly only possible for part-time colleagues. You may be able to get some data through freedom of information requests; but if you go down this latter route, do it sensibly, not vexatiously. Case studies can be related to a single project, e.g. a flagship development, a framework agreement, infrastructure for one-off sporting events, or alternatively a single event at some point in time, e.g. a one-off financial crisis, or liquidation of a single company, or a landmark legal dispute.

Surveys embrace a wide variety of approaches to collecting data. Surveys are often associated with questionnaires of some kind, usually, but not always, involving people. They may be surveys of buildings. Surveys may imply an intent to collect data from a wider audience. In construction, surveys of single plots of land or single buildings perhaps fall more easily into the case study category. However, they can become surveys of a population if there are larger numbers involved, for example, accessibility surveys of public buildings. Similarly focus groups may fall between being a case study or, if used often, a survey. A study that mostly analyses secondary data from archives or the Internet becomes a survey of relevant literature.

Construction presents many opportunities to undertake laboratory or fieldwork experiments. Land surveying, materials technology and science, and environmental science are parts of the built environment curriculum that can use experiments as a way of collecting data. Various concrete mix designs can be trialled to test properties of concrete such as compressive strength, density, and water absorption. The design mixes may vary by type and volume of aggregates, cement, water content and plasticiser. Experiments on people are often undertaken in other disciplines, perhaps with a control group (that carries on as normal) and an experimental group (that has some variable imposed upon it). A construction company may trial a new piece of technology on some sites. Some parallel sites continue using old systems as the control group, and the trialled sites become the experimental groups; some degree of success is then measured across both groups to see if there is improvement.

Ethnography involves you immersing yourself in the life of participants to attain a cultural understanding of their environment. You may wish to share in their experiences and, detaching yourself from inherent bias, see things from their perspective. Construction managers may seek cultural understandings of craft workers, particularly in the context of health and safety. You may, for example, take time out from normal everyday site activities to get closer to craftspeople than would normally be the case.

Grounded theory lends itself to collection of qualitative data through unstructured or semi-structured interviews. There are no pre-conceived ideas; the term 'grounded' is used to denote new theories that may start to emerge from the data. The process is inductive (bottom up, from data collection to the theory) rather than deductive (top down, from the theory to the data collection). These theories may be developed into hypotheses from the data for subsequent testing in other studies.

Narrative enquiries also lend themselves to collection of data by interviews, though the intent is that respondents will tell of their stories, rather than a semi-structured or structured format. The data collection process may need to be over a longer period of time than that normally associated with interviews. You would need to be respectful of the time of interviewees. Most often individuals will not have time to sit in several interviews over an hour or two each; however, you may find situations where this is possible. For example, you may find willing professionals who are between projects, or serving notice at one employer before moving to another, or towards the end of their careers.

## 5.3 Types of data

### Primary or secondary data

When setting objectives, it is reasonable that you are mindful of limitations in getting data. You cannot be too ambitious and aim for objectives that can only be met by data that are likely to be confidential, or that involves concepts that are too difficult to measure.

Data may be loosely classified as primary or secondary. The boundaries between the two can be blurred. Primary data can be defined as new data generated by the efforts of the researcher; that is, the words or numbers that you use in your analysis are created specifically for your research. Examples of primary data are: questionnaires, individual interviews, focus groups, bespoke laboratory experiments, action research or participation.

Secondary data are existing data. It may be recent, or historic and perhaps in archives. Often it is published in the public domain, either paper based or electronically. The possibilities to collect secondary data are endless. The European Union publishes a whole host of statistical data. Data are available on economic performance, cost, environmental pollution, traffic flows, and safety. Other sources include the government statistical office, government quangos, local authorities, professional institutions, company reports, and price books. Data may be taken from the published work of other researchers, provided that it is acknowledged and cited. The thrust of such an approach could be to create some originality by analysing the data in a different sort of way, or perhaps bringing together data from two different sources for analysis. Secondary data may be company specific, such as production data that have been collected routinely to measure some aspect of company performance. Other examples include costs, safety targets, personnel records, snag or defect sheets, etc. Part-time students may be in an excellent position to make use of data from their workplaces, providing permissions are obtained and anonymity protected.

Students often prefer the primary route as a quick and easy exercise in data grab. The secondary data route requires perhaps a little more thought, insight and ingenuity. A secondary approach may involve collecting data about lots of variables within a given problem area, and then 'poking around' those data to look for relationships or causes and effects. If you are measuring two variables, it may be appropriate to use primary data for one and secondary for the other; or of course primary for both or secondary for both. Whichever type of data you use should be driven by the objective.

### Objective or subjective data; hard or soft

Objective data are defined as 'actually existing, real ... facts uncoloured by feelings or opinions ...'. Subjective data are defined as 'a person's views ... proceeding from personal ...

individuality. . .'. Do not confuse 'objective' in the context of a 'study objective' and 'objective data'. It may help to think of objective data as 'hard' data, and subjective data as 'soft'. Lots of quantitative or statistical data are objective or hard. Scientific measures around heat, light, sound, strengths of materials, and densities are also 'hard'. Opinion surveys can be classified as subjective or soft data. On the one hand, it can be argued that just because people are of the opinion that product 'A' is better than product 'B', it does not mean that this is the case in a technical sense. However, people's opinions influence their behaviours and purchasing decisions; these types of decisions can be drivers in the economy, and it is important that we measure them.

The boundaries between subjectivity and objectivity can be blurred. The 'soft' opinions expressed qualitatively by respondents in closed questions are converted to numbers and treated as though hard data. In your dissertation, you need to make a judgement about where your data lie on the continuum between the two. This judgement will underpin assertions that you will also make about the validity of your research. Oppenheim (1992, p.179) distinguishes between opinions, attitudes and values. Opinions may be fairly loose and people may be persuaded to change them. Attitudes are more deep-rooted; it is more difficult to change attitude than to change opinion. Values are more deep rooted still and are only likely to change over a long time span. It may be argued that data at the 'values' end of the continuum are best measured qualitatively.

Whether you should strive to use objective data or subjective data in your dissertation should be driven by your study objective. Businesses need to know and measure how customers, employees and other stakeholders are judging them subjectively. If your objective is related to judgements about issues such as job or product or systems satisfaction, or aesthetic appeal, you should use subjective measures. Measuring knowledge of professionals, perhaps in the area of sustainability, is arguably subjective.

Some research may use a basket, comprising subjective and objective measures. Subjective measures will be used where appropriate, but to improve the validity of the measures, wherever possible, objective data will be used. Part of a piloting process may involve trying to move subjective measures to objective. For example, quality of life is a concept measured universally. It is an important issue for society, but it also has its applications to construction. There are independent comparative measures of quality of life in different countries. The UK Government, through the Audit Commission (2007), measures it in local communities. Its published initiative is called 'Local quality of life indicators - supporting local communities to become sustainable'. The areas measured are: people and place, community cohesion and involvement, community safety, culture and leisure, economic well-being, education and life-long learning, environment, health and social well-being, housing, transport and access. One measure of culture and leisure is subjective; 'the percentage of residents who think that their local area, over the past three years, has got better or stayed the same'. One measure of environment is objective; 'the volume of household waste collected and the proportion recycled'.

You may wish to measure some part of quality of life in organisations. You may be able to base your measure on a subjective opinion survey, and justify this on the basis that quality of lives is 'in the eyes of the beholder'. A questionnaire survey could reasonably be sent to employees in construction companies to measure their perceptions about quality of life.

You may make the judgement that your study objective is best met by hard data. However, given that you are limited by your time and your ability to get access to data (some of which may be confidential), you may be able to justify using soft data in your methodology. The issue of justification is important, as is an acknowledgment in the 'limitations and criticisms of the study' about how using soft measures have limited the validity of the study. If you cover these latter two items well, there is no reason why your dissertation should attract less marks than a dissertation that has been able to use hard data. For example, take a study involving material waste. You may wish to measure the amount of waste on a number of sites. It may be limited to (the population), work carried out directly by main contractors, or alternatively specialists. Your definition of waste may be relatively broad; it draws on an authoritative definition, but includes waste caused by design mistakes, variations, cutting, theft, over-ordering, breakage before use, and damage after fixing. It does not include other elements of waste, such as time wasted by designers, managers, craftspeople or operatives. You do not want to measure waste by number of skips used on site, since that measure includes all trades. You decide that the objective measure is to determine from material invoices, the quantities of materials paid for. This will be compared with quantities measured from drawings. These will be accurate re-measures, taking account of variations, similar to those often used in final accounts. Part-time students may be able to get at these data in their own organisations for a reasonably large sample size. However, full-time students may find it impossible. There is the possibility that some element of material waste could be measured by a clip board survey, walking around sites and making observations about materials waste. This too is impossible for full-time students. Alternatively, therefore, full-time students may design a questionnaire. The example below shows a basket of questions administered to site foreman, with possible responses:

- How often do you check material deliveries for quality?
- Never/not very often/sometimes/often/always
- How often do you check material deliveries for correct quantity delivered?
- Never/not very often/sometimes/often/always
- How often do you check materials whilst they are in storage?
- Never/not very often/sometimes/often/always
- How important do you think it is to give tradespeople 'designed cutting lists' to minimise cutting waste?
- Very important/important/possibly important/probably not important/definitely not important
- How important is it to protect vulnerable materials on site after installation?
- Very important/important/possibly important/probably not important/definitely not important

On the one hand, you may wish to argue that these questions give you a 'measure' of material waste. However, such an argument is tenuous; these questions are not a valid measure of waste. Therefore, it may be better to change your objective, such that you do not seek to measure waste in its hard or objective form. Better that you change your definition of waste. Perhaps the five questions are subjectively measuring 'care in controlling waste'. You should articulate in your introduction or methodology that your intent was to measure waste, but recognising your constraints, you had to change this to 'care in controlling waste'. If your objective has a variable which can reasonably be measured using hard data, you should design your methodology around that measure.

## 5.4 Questionnaires

Some people argue that questionnaires are used too frequently by students. They are a convenient way to 'grab' some data. If students are busy, questionnaires are quick to administer and they allow students to take an easy passive approach to data collection rather than the more difficult proactive approach. Some people become extremely cynical about questionnaires, to the extent that they are the subject of severe criticism. Industry can become weary of receiving questionnaires from students in local universities; there is a need to respect time constraints on professional people and others. The judgement to be made about whether questionnaires are most appropriate is 'is this the best way to meet the study objectives?'. If the answer is 'no', do not use one. If the answer is 'yes', you should use one, and be prepared to robustly justify its use in your methodology chapter. When external examiners review dissertations at the end of the academic year, they should reasonably expect to see a range of data collection methods used by students. There is something wrong if most students use a questionnaire, but it would also be unusual if none do. To re-iterate, if the best way to meet the study objectives is a questionnaire, do it.

There are many legitimate concerns about the validity of questionnaires as a method of collecting data. Validity is defined here as 'how well does a questionnaire really measure what it purports to measure'. Biased data may result from low response rates, e.g. 30 of 100 questionnaires returned; what would the 70 respondents have said if they had completed the questionnaire? Therefore, even if the numbers in statistical analysis indicate significant results, they may be of dubious value. Consider two examples: (a) a questionnaire survey to social housing tenants to measure the quality of service provided by landlords in dealing with maintenance issues; there is a possibility that the tenants who will have the greatest propensity to reply are those who are dissatisfied with the service, and therefore scores may be poor; (b) a questionnaire survey to measure commitment to health and safety will most likely be responded to by companies that are highly committed, and therefore scores will be good. In both cases, the results do not reflect the population; they are biased.

A questionnaire is used to 'measure' something within a defined population. Careful selection of the sample must take place. It is very useful to keep focused on the concept of 'measurement'. These measures are often about people's opinions, values, knowledge or behaviours. Questionnaires that are not administered face-to-face usually have closed questions with a choice of pre-selected answers. There may be a few open questions that require respondents to express their own words qualitatively.

Students sometimes mistakenly use questionnaires to seek answers to interesting questions not directly related to study objectives. Do not stray from the study objectives and the variables in those objectives that you want to measure. The questionnaire may have few or no questions; it may be based on given statements. Respondents are required to indicate their agreement or disagreement to statements; scores are developed on the Likert scale, developed by the American Psychologist Rensis Likert (1903–1981). An alternative to 'agree/disagree' is 'approve/disapprove'.

There can be two parts to questionnaires: (a) general questions to measure subject variables, and (b) measurement questions. Think about whether you prefer general questions at the start or end of the questionnaire. There is no definitive way; general questions tend to be easily answered, and respondents may view them as an easy lead-in to the questionnaire.

In a quantitative questionnaire it can be very fruitful to include a last question: 'please add any comments you may have below'. Such qualitative responses may give you new insights that you can use to flavour your discussion and conclusions. If your questionnaire is interesting, topical and has reached the correct people or population, you should get lots of comments to help you. Some respondents may 'pour their hearts out to you' about how passionately they feel about the topic area of your study.

Type these comments up verbatim, and include them in an appendix. At the end of your questionnaire always say thank you, and invite respondents to give you their contact details if they would like a summary of your findings. Always ensure that you offer and protect the confidentiality and anonymity of respondents.

### Piloting the questionnaire

Piloting a questionnaire helps to improve its validity. A questionnaire that you have designed may be perfectly clear to you, but not clear to readers. When you send the questionnaire out, this is an important milestone in your study; no turning back. If the questionnaire contains a fundamental error or is poor, the rest of the study will most likely be poor. You will not have time to re-administer a poor questionnaire. Work with your supervisor on several iterations. When you are both happy, start the piloting process. As a starting point, select three or four people that ideally represent your population, for part-time students perhaps colleagues at work. Sit with them whilst they complete the questionnaire. Remain silent, since you will not be able to speak to respondents when the questionnaire goes live. Observe them whilst they complete the document, and time how long it takes. After they have completed it, have an informal feedback session:

- How did they find it?
- Were the instructions clear?
- Were the questions clear: if not, which ones?
- If they paused over any questions, why?
- What do they think of the scales?
- Did they object to answering any questions?
- Was the layout attractive?
- Do they think the questions are good measures of the variables?
- Do they have any other comments or suggestions for improvement?

The second part of the piloting process is to administer the questionnaire to say 10% of the sample. There is potentially a huge constraint on you here; time. To send out a questionnaire, get the data back, do some analysis and make changes before administering it again to the whole sample, is potentially very time consuming. Make sure you allow for it in your programme. Not all students do so. You may 'target' the questionnaire to help get best response rate, recognising that the data set is arguably biased. That is no matter though, on the basis that it will not be used in the main study analysis. Tell respondents it is part of a piloting process, and at the end of the questionnaire, ask the same questions you have asked to people who gave you feedback face-to-face. You should tabulate results, and proceed to analyse the data using appropriate analytical methods. Ideally go on also to formulate interim findings, conclusions and recommendations. Chapter 8 details some possible statistical checks for internal reliability of questions. Check to see if results cluster around the central point. Any faults in the research questionnaire should be corrected before it is issued to the remainder of the sample. Explain clearly in your method chapter your piloting process, and justify any changes made to your questionnaire between the pilot and main study.

## Coding questionnaires

Table 4.3 is adapted to form Table 5.1. The variable being measured is commitment of speculative housing bricklayers to producing high quality work. Table 5.1 illustrates how scores and codes can be used for subject variables to summarise the data set. Measurement questions are best 'scored' with numbers starting at zero, thus: Q1 'I make sure that my tools have been cleaned after use'. Always = 4, usually = 3, sometimes = 2, rarely = 1, never = 0. Q2 to Q8 (see Chapter 4) are scored similarly. General questions or subject variables are best 'coded' with numbers, rather than scored, and whilst a 0 code is possible, it is more usual to start at '1', e.g. a question about gender may code male as '1', and female as '2' (or the other way round). Ages of respondents may be coded as 18–25 years = 1, 26–37 years = 2, 38–55 years = 3 and > 55 years = 4. The single number codes are used as a substitute for words (or age ranges), numbers lending themselves more readily to spreadsheets. The numbers codes for general questions are arbitrary and have no inherent value. If you have a batch of completed questionnaires, coding with numbers allows responses to be summarised in one single spreadsheet. The only subject variable shown in Table 5.1 is age, though others are possible such as size of company the bricklayer works for, years' experience as a bricklayer, and average weekly pay. Since there are four groups for the subject variable age, it is difficult to do meaningful analysis around this variable if a small sample size is split into four much smaller groups. Therefore, in the last column of the table, the age of bricklayers is split into two less small groups. You should browse the data so that the two groups will have as near as possible equal numbers in them. In this case a group reclassified with the ranges 18–55 years = 1 will have four bricklayers, and a group reclassified as > 55 years = 2 has six bricklayers. The mean score for group 1 is 40.7% and for group 2, 72.1%. These results warrant further investigation and discussion in later stages of the dissertation, including perhaps some qualitative work with bricklayers or their supervisors.

For questions that offer closed tick-box responses, be clear in your instruction about whether respondents are required to tick only one box, or give multiple responses. For example, if you ask about respondent qualifications, and offer a number of choices (perhaps linked to the UK census categories), you may only need one tick against the highest level qualification. If you really only need one tick, it is arguably best to ask for that. Alternatively, you may really need in the context of your study respondents to tick more than one box. If you do, to give instructions to respondents, use the language of an authoritative source, such as 'tick all the qualifications that apply or, if not specified, the nearest equivalent' (Office for National Statistics 2001). The format for coding and recording multiple responses is a little more cumbersome, though that should not deter you. Table 5.2 summarises a hypothetical data set, where the code 0 is used to indicate that the respondent does not have that qualification, and the code 1 to indicate that the respondent does have it. Using these numbers, a spread sheet will give the total number in each category. Each question is coded

**Table 5.1** Commitment of speculative housing bricklayers to quality, including the subject variable age.

| | Q1 | Q2 | Q3 | Q4 | Q5 | Q6 | Q7 | Q8 | Q1 to Q8 total | % score for commitment | Age of bricklayer; 18–25 years = 1, 26–37 years = 2, 38–55 years = 3 and >55 years = 4 | Age of bricklayer; 18–55 years = 1, >55 years = 2 | Score group 1; 18–55 years | Score group 2; >55 years |
|---|---|---|---|---|---|---|---|---|---|---|---|---|---|---|
| R1 | 4 | 4 | 2 | 3 | 2 | 3 | 2 | 3 | 23 | 72 | 4 | 2 | | 72 |
| R2 | 4 | 3 | 3 | 2 | 2 | 3 | 1 | 2 | 20 | 63 | 4 | 2 | | 63 |
| R3 | 4 | 3 | 2 | 3 | 1 | 3 | 1 | 4 | 20 | 63 | 4 | 2 | | 63 |
| R4 | 3 | 2 | 2 | 3 | 1 | 1 | 1 | 2 | 15 | 47 | 3 | 1 | 47 | |
| R5 | 4 | 3 | 4 | 3 | 1 | 3 | 0 | 2 | 20 | 63 | 4 | 2 | | 63 |
| R6 | 4 | 2 | 3 | 3 | 2 | 3 | 0 | 3 | 20 | 63 | 2 | 1 | 63 | |
| R7 | 4 | 4 | 3 | 2 | 2 | 3 | 1 | 3 | 23 | 72 | 4 | 2 | | 72 |
| R8 | 3 | 2 | 2 | 2 | 2 | 2 | 2 | 2 | 17 | 53 | 2 | 1 | 53 | |
| R9 | 0 | 0 | 0 | 0 | 0 | 0 | 0 | 0 | 0 | 0 | 1 | 1 | 0 | |
| R10 | 4 | 4 | 4 | 4 | 4 | 4 | 4 | 4 | 32 | 100 | 4 | 2 | | 100 |
| Totals | 34 | 27 | 25 | 26 | 17 | 24 | 12 | 25 | | 596 | | | 163 | 433 |
| n | | | | | | | | | | | 10 | | 4 | 6 |
| Mean | 3.4 | 2.7 | 2.5 | 2.6 | 1.7 | 2.4 | 1.2 | 2.5 | | 59.6 say 60% | | | 40.7 | 72.1 |

**Table 5.2** Data set in a spreadsheet for respondent qualifications.

| | Q1a Craft | Q1b A levels | Q1c NVQ | Q1d HNC/D | Q1e Degree | Q1f Professional | Q1g Postgraduate |
|---|---|---|---|---|---|---|---|
| R1 | 1 | 0 | 0 | 1 | 0 | 1 | 0 |
| R2 | 1 | 0 | 0 | 1 | 0 | 0 | 0 |
| R3 | 0 | 1 | 0 | 0 | 1 | 1 | 1 |
| R4 | 0 | 0 | 0 | 1 | 1 | 1 | 0 |
| Totals | 2 | 1 | 0 | 3 | 2 | 2 | 1 |

with the letter a, b, c, etc. Since this one question requires seven columns in a spreadsheet, it is likely that you will need more than one sheet of A4 in your dissertation for your total data set, or a very carefully folded sheet of A3. It is possible that you could ask for multiple ticks, and this leaves you the option of sorting for presentation in the spreadsheet and subsequent analysis, depending on the range of answers received. The questions for the data set in Table 5.2 are:

Q1. What qualifications do you have? Tick all the qualifications that apply or, if not specified, the nearest equivalent.

Craft qualification                                                            ☐
2 + A levels                                                                   ☐
NVQ level 4–5                                                                  ☐
HNC, HND                                                                       ☐
First degree or equivalent                                                     ☐
Professional qualifications (e.g. MCIOB, MRICS, MCIAT, MRIBA)                   ☐
Postgraduate qualification                                                     ☐

## A basket of questions to measure variables or multiple item scales

If the purpose of a questionnaire is to measure something, it should measure it really well; that is, to measure it in depth, not superficially. The thing that is being measured is at least one variable in an objective. The questionnaire may seek to measure two or more variables. If this is the case, you should be absolutely clear which questions are contributing towards the measurement of which variable. To measure something really well and in depth, you should use a basket of questions or provide a basket of statements that are all related to the variable you are measuring.

To ask just one question to measure a variable would be a nonsense. Take, for example, a questionnaire to measure the variable 'compliance of employees with company quality procedures'. This 'could' be measured with one question such as 'how often do you adhere to company quality procedures?' There could be five possible responses given: 'never, rarely, sometimes, usually, always'. Each of these labels could be given a number, 0 to 4. Respondents may quickly (anecdotally) give responses. A score of 3 out of 4 for one individual becomes 75% (0 = 0%, 1 = 25%, 2 = 50%, 3 = 75%, and 4 = 100%) and some statistical analysis could follow. This measure of 75% is an extremely poor measure; it is superficial and not valid.

To get at a better measurement of 'compliance with company quality procedures', you must ask a basket of questions. You must work hard to tease the measurement out from respondents.

An example, is the *Sunday Times* survey of the best companies to work for. The survey is repeated each year. Construction companies are rightly proud if they achieve a ranking in the top 100. The survey measures one key variable 'how good is the company to work for'. If it were just to ask the single question 'how good is your company to work for', that would be a poor measure. To measure this one variable, the survey gives 66 statements, to which respondents can indicate agreement or disagreement over a seven point scale (strongly disagree, disagree, slightly agree, neither agree nor disagree, slightly agree, agree, strongly agree). Each answer can be given a number 0–6 (note not 1–7), with a high score indicating a good company to work for. The minimum score for a company, though unlikely, is zero, and the maximum score is 396 (66 × 6). For simplicity, total scores can be expressed as a percentage, such that 198 points out of 396 = 50%.

Within a basket of questions or statements, the 'poles' for questions should be switched to avoid the consequence of yea-sayers or nay-sayers. Table 5.3 shows just two of the 66 *Sunday Times* questions, which are at opposite poles, and with the researcher's codes shown. Employees that think their companies are best will tick on opposite sides of the paper, strongly agree for Q1, and strongly disagree for Q2. If questions were not switched, after answering four or five questions, respondents would quickly grasp that ticks for a best company will always be on one side of the paper, and for a bad company on the other side. They may become flippant or bored in answering questions and merely tick statements in the same vertical column without giving responses too much thought. Since there are 66 questions, it cannot be expected that respondents will labour long and hard over each response, but it can be expected that they pause momentarily, and digest each question. If questions are switched, this slows respondents down a little, and subconsciously forces them to think before answering. There should not be a structure to the switching, e.g. five questions in one direction and then five in the next; it should be thoughtful but fairly random.

The width of measurement scales for each individual question or statement should also be given careful thought. A mere two point wide answer scale is possible, e.g. 'do you comply with company quality procedures?' with two categorical answers as tick box options, 'yes'

**Table 5.3** Two questions, as examples, from the sixty-six questions asked, in the *Sunday Times* survey of the 100 best companies to work for. Source: *Sunday Times* (2009).

| Question | Strongly disagree | Disagree | Slightly disagree | Neither agree nor disagree | Slightly agree | Agree | Strongly agree |
|---|---|---|---|---|---|---|---|
| Q1; the leader of this company is full of positive energy | 0 | 1 | 2 | 3 | 4 | 5 | 6 |
| Q2; I am under so much pressure at work I can't concentrate | 6 | 5 | 4 | 3 | 2 | 1 | 0 |

and 'no'. In the context that you wish to attain really good measurements and maximise the potential for robust analysis, it is not appropriate to limit answers so narrowly. Life is often more complex than merely 'yes' or 'no', or 'black or white' responses. You need to tease out and measure the 'maybe' and 'grey' possibilities. At the other extreme why not a 10 point scale (1–10), or 11 point (0–10), or even wider? Ten point scales are often used; respondents may or may not be quite as comfortable to place their judgements in this relatively wide scale. It may be easier for respondents with a narrower scale; five or six is probably quite easy; at seven it arguably starts to get tough.

Whether to offer respondents an odd or even scale should be given careful thought. An odd number allows respondents to take the easy middle option and be neutral in their answers; 'sit on the fence'. An even number forces respondents onto one side or the other. Fink (1995, p.53) argues that current thinking suggests that 5–7 point scales are adequate for most surveys that generate ordinal data, and there is no conclusive evidence supporting odd or even scales; it depends on the survey's needs.

It is usual to give respondents only a 'word' or 'statement' to tick, without a number. The number is not seen by the respondent and only allocated by the researcher later. On a wide scale though of 0–10, sometimes only numbers are given, or a mixture of words and numbers. There is an argument that to improve validity, numbers must have a descriptor. What a number means to one person may not be what a number means to another. Take for example 7 out of 10. On an undergraduate degree programme, 7 out of 10 is an 'excellent' score at first class honours standard. However, 'excellent' in Constructing Excellence Key Performance Indicators is thought by some to be only achieved at 9 or 10 out of 10. To write descriptors for all 11 points on a 0–10 scale can be difficult. A compromise position is to write labels to cover a narrow range of numbers, e.g. 0–3 = poor, 4–5 = satisfactory, 6–7 = good, 8–9 = very good, 10 = excellent. Respondents who may wish to distinguish between say 6 and 7 'good', may have in mind a judgement at the lower or upper end of 'good'. School teachers routinely use upper and lower scores, e.g. B + is clearly grade B, and not quite so good as −A, and −A is clearly an A; or is it?

Paper-based questionnaires can be administered face-to-face, or handed to people personally so that they can complete them in their own time. They can be sent out by post, but if you do this be sure to include a stamped self-addressed envelope. Be sure that the layout of your questionnaire is attractive, and use tick boxes to allow respondents to complete them quickly. On-line web-based electronic questionnaires are more convenient and cost effective for researchers, and are probably also convenient for respondents. There is a lot of commercial software easily available for free student use over a limited time period of say one month. Appropriate key words in web search engines will locate alternative software. You can design your own survey. Tick boxes allow category or quantitative responses. Respondents can also type in their qualitative responses, if required. There are problems about finding 'names' to send your questionnaire to. It is usual to send an e-mail with a web address for respondents to click onto, and 'submit' their answers at the end. In some organisations, e-mail addresses are freely available. In other organisations, they are not. If addresses are not available, it may only be possible to send your e-mail request to a generic business address, and ask the recipient to forward it to relevant colleagues. If you use electronic surveys, you must acknowledge in your dissertation their potential weaknesses, such as, 'do they implicitly give you a biased sample'? You can justify the use of electronic questionnaires, even though postal surveys may be more valid, just on the basis of cost. Postage and printing costs can be expensive. Your university will not want you to incur undue cost; it is just for you to recognise and write about the weaknesses.

Some things to avoid in questionnaires:

(a)  Loaded questions, such as do you think method A is better than method B; you must use a neutral format.

(b)  Double questions that often contain the word 'and', such as 'do you have responsibility and authority in your workplace'; this one question should be written as two separate questions.

(c)  Sensitive personal questions; people may not wish to answer questions about income, religious beliefs, alcohol consumption, or precise age.

(d)  Vague questions; 'how would you describe your health?' is better written as 'how would you describe your health in the last three months?'.

(e)  Limiting closed responses, e.g. age bands, such as the top range for professionals is 55–65; better to say over 55.

(f)  Questions that require respondents to reach into files for data; if this is the case non-response rates will be very high—most questionnaires should allow respondents to complete them reasonably quickly, such that it is 'tick, tick, tick, some qualitative comments, finished and submit'.

(g)  Time-consuming questionnaires to busy business people—perhaps 10 minutes maximum.

(h)  Skip patterns: a skip pattern anticipates that some questions will not apply to all respondents, e.g. if answered 'yes' to this question, please go to question … Do not attempt to ask for sensitive business information, such as safety statistics, cost data, or anything else that businesses may consider confidential.

Some things you should do in questionnaires:

(a)  Be conversational in your tone.

(b)  Consider the target audience, e.g. questions to young people should be different to those to executives.

(c)  Make sure any closed responses to questions are exhausted; this may mean including options such as, 'don't know', 'unsure', 'uncertain', 'neither agree nor disagree'.

(d)  Use tick boxes or table structures where possible to facilitate speedy completion.

(e)  Assure respondents that data provided will be kept confidential—and make sure you comply with that.

(f)  Assure respondents that their anonymity will be protected.

(g)  Offer respondents a summary of your results; you will need to ask for contact details if they indicate they would like them.

(h)  Always finish with a 'thank you for completing the questionnaire'.

Include with your questionnaire a covering letter if by post, or a covering statement if electronic. The statement may include one sentence to detail each of the following:

(a)  The context of the questionnaire (for your dissertation).

(b)  The title or objective or a description of the problem.

(c)  Probable time to complete, e.g. 'you should be able to complete this questionnaire in less than 10 minutes'.

(d)  The date for return of the questionnaire; if you set this date too distant, it may get forgotten about, if it is too short people may not have time—perhaps 2 weeks is appropriate?

Select all these words carefully since you are trying to entice respondents. If the survey is electronic, potential respondents may make one of two instant judgements, having read your statement; that is, either to click onto to the web link to open the questionnaire, or to delete.

You will need to predict a response rate to your questionnaire to ensure that you have a sufficient number of responses for analysis. Postal survey response rates may be as low as 25%, so if you need 30 responses, 120 should be distributed. A more optimistic response rate may be predicted for electronic surveys, but whatever the response rate is depends upon many variables, e.g. relevance of the topic area to the sample selected, attractiveness of design, and time for respondents to complete.

For each question or statement, only as a last resort should you use your own questions. It is likely that, for many concepts you wish to measure, a measurement tool has already been designed. You should be seeking to build from a platform created by others, not reinventing the wheel. So you should use what has been used elsewhere, providing it is legitimately in the public domain, and providing that you cite the source in your methodology chapter. You may wish to cite the source on the questionnaire itself, although for student work perhaps that is not absolutely necessary. Your questionnaire may be a collection of questions or statements from a variety of sources; if so, cite each source. You may take some questions or statements and adapt them to the context of your study; if so, in your methodology chapter cite the original source and justify adaptations you have made. You must be sure not to infringe copyright, since some questionnaires may only be used by payment of a fee. Whilst the most valid questionnaires are those already in the public domain, do not let that deter you from thinking innovatively about new questions. If you do this, the argument that questions in the public domain are proved to be more valid by regular use can be compensated by robust piloting of your questions, and by doing tests of internal reliability (see Chapter 8).

### Using a basket of questions in ordinal closed-response scales

Douglas McGregor (1906–1964) used a basket of questions to measure management style. Theory 'X' was an authoritarian management style; theory 'Y' was a participative style. He suggested that enlightened managers are theory 'Y' and get better results. To measure management style, he used a basket of questions, based upon assumptions of managers about workers. The basket of questions is a better measure of management style than the single question 'what is your management style?'

The 'X-Y Theory' Questionnaire, was proposed by McGregor (1960), in his text *The Human Side of Enterprise,* thus (Business 2010):

Score the statements (5 = always, 4 = mostly, 3 = often, 2 = occasionally, 1 = rarely, 0 = never)

To indicate whether the situation and management style is 'X' or 'Y':

(1) My boss asks me politely to do things, gives me reasons why, and invites my suggestions.
(2) I am encouraged to learn skills outside of my immediate area of responsibility.
(3) I am left to work without interference from my boss, but help is available if I want it.
(4) I am given credit and praise when I do good work or put in extra effort.
(5) People leaving the company are given an 'exit interview' to hear their views on the organisation.
(6) I am incentivised to work hard and well.
(7) If I want extra responsibility, my boss will find a way to give it to me.
(8) If I want extra training, my boss will help me find how to get it or will arrange it.
(9) I call my boss and my boss's boss by their first names.
(10) My boss is available for me to discuss my concerns or worries or suggestions.
(11) I know what the company's aims and targets are.
(12) I am told how the company is performing on a regular basis.
(13) I am given an opportunity to solve problems connected with my work.
(14) My boss tells me what is happening in the organisation.
(15) I have regular meetings with my boss to discuss how I can improve and develop.

Since there are 15 questions with a maximum score of 5 each, the total score maximum is 75. The qualitative labels against the range of scores are:

60–75 = strong Y-theory management (effective short and long term)
45–59 = generally Y-theory management
16–44 = generally X-theory management
0–5 = strongly X-theory management (autocratic, may be effective short-term, poor long-term)

These questions are no doubt often used and validated in many research studies. As a measurement tool, provided it is administered appropriately, it may be argued it is valid. If you were to measure management style, inferential statistical analysis would be possible if you also measured some other variable, perhaps worker satisfaction, in the context of a problem that you have described.

If you are using the questionnaire to measure two variables in an objective so that you can do inferential statistical analysis, it is important to ask a basket of questions and measure each variable independently of the other. If you ask respondents to make a judgement about a possible link between the two variables (using only one basket of questions), that leaves you with the possibility of only doing descriptive analysis. If you adopt this latter course, on the one hand this is fine, but do recognise that it limits your opportunities to demonstrate your analytical skills. Consider, for example, an objective 'to determine if the approach by clients and contractors to partnering (independent variable [IV]) influences the success of projects (dependent variable [DV])' (Brook and Farrell 2004). The style of statements is switched to prevent the effect of 'yeasayers' and 'naysayers'. Two separate baskets may be:

(1) To measure the IV—approach by clients and contractors
- Consideration was given to the progress and stage of construction when making variations
- The partners appreciated the problems/complexity of the project
- The client was very much part of the team
- All disputes were resolved in an amicable manner
- Partners frequently criticised each other
- A high level of trust existed between the client's and the contractor's teams
- A blame culture was prevalent on the project
- The mutual objectives of the project were clear to all parties

(2) To measure the DV—success of projects
- The project was completed on time
- The level of defects/snagging at practical completion was unacceptable
- The project was a financial success for the contractor
- The client was very satisfied with the quality of the finished product
- The project was completed within the client's budget
- The safety record of the project was poor
- The contractor revisited the project after practical completion on numerous occasions to attend to defects
- The relationship between the client and contractor was sometimes poor during the construction phase

These two separate baskets give you the opportunity to score each variable separately and plot the scores on a scatter diagram; as a researcher you may determine whether there is a link between the variables by undertaking some statistical inferential statistics, e.g. correlation calculations.

Alternatively, consider the use of only one basket of statements; those for the IV basket above are reworded. The instruction to respondents is 'please indicate to what extent you Strongly Agree, Agree, Neither Agree or Disagree, Disagree, or Strongly Disagree with the following statements (success is defined as . . .):

- Whether consideration is given to the progress and stage of construction when making variations influences the success of projects
- Whether partners appreciate the problems/complexity of projects influences the success of projects
- Whether the client is very much part of the team influences the success of projects
- Whether disputes are resolved in an amicable manner influences the success of projects
- Whether partners frequently criticised each other influences the success of projects
- Whether a high level of trust exists between client's and contractor's teams influences the success of projects
- Whether a blame culture is prevalent on projects influences the success of projects
- Whether the mutual objectives of the project are clear to all parties influences the success of projects

Given only one basket of statements, respondents make the link between the variables. As a researcher you may only present descriptive results in tables and histograms.

## Other possible responses in ordinal closed scales

There are lots of possible responses that should be driven by the questions asked. A six point wide scale of satisfaction (completely satisfied/very satisfied/somewhat satisfied/somewhat dissatisfied/very dissatisfied/completely dissatisfied), can alternatively be presented as five or four point wide (completely satisfied/somewhat satisfied/somewhat dissatisfied/completely dissatisfied). Other possible responses are:

- Importance: definitely unimportant/probably unimportant/probably important/definitely important/no opinion/don't know
- Agreement: definitely true/true/don't know/false/definitely false, or definitely yes/ probably yes/probably no/definitely no/uncertain or no opinion
- Frequency: always/very often/fairly often/sometimes/almost never/never, or always/ frequently/sometimes/rarely/never
- Intensity: none/very mild/mild/moderate/severe, or very much/much/a fair amount/ a little/not at all
- Influence: big problem/moderate problem/small problem/very small problem/no problem
- Comparison: much more than others/somewhat more than others/about the same as others/somewhat less than others/much less than others.

The centre point of a five-wide Likert scale below is labelled 'uncertain'. Alternatively, it could be 'undecided' or 'neither agree or disagree':

- Strongly agree/agree/*uncertain*/disagree/strongly disagree

A seven point scale for Likert could be:

- Strongly agree/agree/slightly agree/*uncertain*/disagree/slightly disagree/strongly disagree

## Ranking studies

The objective of a study may be to ascertain ranking judgements of respondents about issues. A list of factors may be given, and you ask respondents to rank them in terms of importance, quality, frequency of occurrence, frequency of use or such like. The example below is a study that seeks to rank respondent opinion about factors that keep sites safe. There are three generic areas: bespoke risk assessments, supervision, and training operatives. Other areas could be selected too as part of a study, such as use of personal protective equipment, training and education for managers, and prevalence of production incentive schemes. Rankings are sought for five items in each generic area, and then mean rankings taken. Table 5.4 shows a small hypothetical data set; note a sample size of five is too small for a dissertation. The question to respondents is 'please rank each of the following factors in order of their importance in keeping sites safe, ranking the most important as 1 and the least important as 5':

**Table 5.4** Rankings of factors that keep sites safe.

| | Questions | | | | | | | | | | | | | |
| | People involved | | | | | Supervisors | | | | | Operatives and craftspeople | | | | |
| | 1a | 1b | 1c | 1d | 1e | 2a | 2b | 2c | 2d | 2e | 3a | 3b | 3c | 3d | 3e |
|---|---|---|---|---|---|---|---|---|---|---|---|---|---|---|---|
| R1 | 1 | 2 | 4 | 5 | 3 | 1 | 2 | 5 | 4 | 3 | 1 | 2 | 3 | 4 | 5 |
| R2 | 2 | 3 | 1 | 4 | 5 | 1 | 2 | 5 | 4 | 3 | 2 | 3 | 5 | 4 | 1 |
| R3 | 1 | 3 | 2 | 4 | 5 | 1 | 2 | 5 | 4 | 3 | 3 | 4 | 1 | 5 | 2 |
| R4 | 2 | 4 | 1 | 5 | 3 | 1 | 2 | 5 | 4 | 3 | 4 | 3 | 2 | 1 | 5 |
| R5 | 1 | 5 | 2 | 3 | 4 | 1 | 2 | 5 | 4 | 3 | 5 | 3 | 4 | 1 | 2 |
| Total | 7 | 17 | 10 | 21 | 20 | 5 | 10 | 25 | 20 | 15 | 15 | 15 | 15 | 15 | 15 |
| Mean ranks | 1.4 | 3.4 | 2 | 4.2 | 4 | 1 | 2 | 5 | 4 | 3 | 3 | 3 | 3 | 3 | 3 |

Bespoke risk assessments

(1)   When writing risk assessments, the people involved should be:

(a) The operative or craftsperson                                    ☐
(b) Trade foreman                                                    ☐
(c) Site manager                                                     ☐
(d) Contracts manager                                               ☐
(e) Safety officer                                                   ☐

Supervision

(2)   Supervisors should be:

(a) Experienced in construction                                      ☐
(b) Experienced in the type of project under construction            ☐
(c) Ex-craftspeople                                                  ☐
(d) Graduates                                                        ☐
(e) Sufficient in number on site                                     ☐

Training operatives

(3)   Operatives and craftspeople become most safety aware by:

(a) Site experience                                                  ☐
(b) Tool box talks                                                   ☐
(c) Off site short courses                                           ☐
(e) Threat of disciplinary action                                    ☐
(f) Safety incentive schemes                                         ☐

For question 1, the lead rank item is 1a, whilst the worst ranked items are 1d and 1e. These mean ranks should promote discussion, including identifying items that should be targeted for action. The data set shown for Q2 is unlikely to occur in practice; there is perfect agreement between the judges, item 2a is unanimously the lead rank, and item 2c the worst ranked. The data set shown for Q3 is again unlikely to occur in practice; there is no agreement between the judges, since the mean ranks for all items 3a to 3e are identical. In practice, whether your mean ranks are close to Q1, Q2 or Q3 should drive the type of discussion.

If you wish to undertake inferential tests for ranking data, the appropriate test is Kendall's coefficient of concordance as explained by either Siegel and Castellan (1988) or Cohen and Holliday (1996). This test will tell you whether differences in rankings between respondents are significant.

Ranking studies lend themselves to items such as causes of delay in projects, causes of material waste, causes of defects or causes of plant breakdown. When providing a list of items to rank, provide enough items in the list to give you the opportunity to get rich data; but not a list of items that is too long, since meaningful ranking may be difficult for respondents. Perhaps ten items to rank is too many and difficult for respondents.

## 5.5 Other analytical tools

The two analytical tools most often described are qualitative and quantitative; the quantitative approach does not have to involve conventional descriptive or inferential statistics. Other tools can be used, which fit only loosely into the qualitative and quantitative domains. Providing they help to improve knowledge and understanding in construction, they are fine. Some examples:

- Drawings or sketches may be used to predict how materials may behave when placed together in composite structures, or how moisture may penetrate into buildings
- Comparison of alternative designs to yield technical or environmentally friendly solutions
- Photographic surveys may be used to compare and contrast material defects in existing properties; use with computer images to enhance visualisation
- Building surveys: compare energy performance certificates across a range of properties or areas
- Critical path programmes or bar charts may illustrate the impact on time of choice of a particular method of construction, perhaps timber frame versus traditional housing, or concrete versus steel multi-storey structures
- Quantity surveyors may wish to take-off parts of a building using Standard Method of Measurement 7th Edition (SMM7), and/or the New Rules of Measurement and/or International Rules of Measurement; comparisons can be made about speed of take off
- Comparison of price estimates may be undertaken based on elemental cost plans, approximate quantities, and composite rates or on different methods of construction
- Land or GPS surveys to compare the accuracy of instrumentation
- Time and motion studies to appraise new methods of working
- Laboratory experiments, such as materials testing, soils testing, behaviour of fluids, heat, light and sound
- Mathematical computations; use of formulae, such as those used in 'U' values, energy calculations or surveying
- Comparison of site investigation reports

- Risk analysis: consider alternative projects in 'what if' models to assess their financial viability, i.e. what if interest rates change, ditto inflation
- Decision trees: compare established theoretical models to a case study situation
- Software programmes: design a measuring tool, and use it to compare two competing products on the market
- Brainstorm alternative building designs to yield best value solution
- Feasibility studies
- Cost/benefit analysis techniques
- Criterion scale judgements; writing qualitative criteria and making judgements against about where actual performance sits on a scale

## 5.6 Incorporating reliability and validity

Reliability and validity are very important and related concepts in research; if you cannot demonstrate that your research is reliable and valid, it is 'worthless'. You should work very hard to maximise reliability and validity. They are not considered in isolated parts of dissertations; assuring validity is not a single act, it is the whole process. You should write at length in the methodology chapter about how you have striven to achieve them. Reliability is part of validity; you cannot have validity without reliability. Reliability is consistency over time. Validity is 'how far a measure really measures the concept that it purports to measure' (Bryman and Cramer 2005, p.80). Therefore it follows that the validity of a whole study is whether it achieves what it purports to achieve; to what extent does it truly meet its objectives. Also be mindful that your objectives should be set to accomplish something worthwhile; modest objectives, but not only to achieve the obvious or something too simplistic.

Reliability is required, for example, with laboratory or surveying equipment. A weigh scale must consistently weigh a 10 kg weight at the same weight, not 10 kg one day, 9.9 kg the next day and 10.1 kg the next day. As part of your study you should report in your dissertation the manufacturers' recommendations for maintenance of equipment, and report on recorded service histories that provide evidence recommendations are being followed. An automatic survey level must consistently give the same difference in height between two fixed points; the difference in height must not change each time the instrument is reset and new readings are taken. You should carry out tests and retests on surveying equipment before collecting data for your dissertation, and report the results of those tests.

A reliable questionnaire would generate the same answers if the survey were repeated. This may suggest that you ask respondents to complete a survey, and then a few weeks later repeat it again with the same respondents. You may then judge the survey reliable if you get the same answers. In theory, this is correct, and you would be able to assert your questions are reliable. But in practice, of course, you cannot usually do this. Reliability may be asserted if the survey has been proved to be successful in many studies elsewhere. If your survey is not well established or if you are designing your own questions, baskets of questions are used to improve the internal reliability of surveys. Chapter 8 illustrates how scatter graphs and correlation coefficients can be used to test internal reliability. The piloting process should be used to detect any questions that are unreliable. If questions are found to be unreliable in the pilot, they should be changed or removed when the main study questionnaire is administered.

The relationship between reliability and validity is often illustrated by drawing an analogy to a clock:

- A clock that measures 61 minutes for one hour, then 60 minutes and then 58 minutes is not reliable; nor is it valid. It cannot be relied upon to measure time
- A clock that measures 61 minutes for one hour, and does that every hour, to the exact second, is highly reliable. We can be sure that for the next hour it will again measure it as though 61 minutes. After the clock has run for 61 minutes, we can definitively say that one hour has just passed. But, it is not valid, since it does not do what it purports to do, that is, measure one hour as 60 minutes
- A clock that measures 60 minutes for one hour, and does that every hour, to the exact second, is highly reliable and valid. We can be sure that for the next hour, it will again measure it as 60 minutes, and that is what it purports to do.

In the example of the weigh scale used above for reliability, a scale that consistently weighs a 10 kg weight at 10.1 kg is reliable, but not valid. We can see from these examples that reliability comes before validity, and we can have reliability without validity. But we cannot have validity without reliability.

You should work to assure the validity of the whole of the study and each individual part; that is done by careful design and pragmatic action. You may also consider the validity of say the measurement part, or the validity of the sampling strategy. The validity of your research holds together like a chain; if one link in the chain is weak, the whole study becomes invalid. You should seek to demonstrate validity in your methodology chapter. Validity first recognises that you are the lead player, and you will self-scrutinise all facets of your study. In the introduction chapter, you must ensure the problem is well founded, has a theoretical base and the variables in objectives are well defined. Take parts of your study to other people who are knowledgeable in the field and ask them to give you feedback. This can be by formal or informal interviews. In the review of the literature you must ensure you are critical and do not miss previous important work. In your methodology, you must select the best method to meet the objective, including the most appropriate population and sample. Large sample size improves validity. Pilot parts of your study, perhaps the data capture and analysis part. Ideally, use a measurement tool that is considered to be the gold standard in your discipline; that is, use something that has been developed elsewhere by authoritative sources (be sure that you cite it), that has been tried and tested by years of experience in respected studies. The analytical method must be robust, rather than using lightweight analytical tools. Towards the end of the document, the discussion and conclusions should be informed by insight, and best developed through dialogue or interviews with experts.

If your study held together reasonably well, but, for example, you were careless in how you picked your sample (one weak link in the chain), or your sample was biased, the whole study is flawed. The consequence of this weak link is that your sample may not reflect the population, and therefore any inference made that results from the sample reflect results as if the whole of the population had been surveyed, is spurious. The analysis that follows will use spurious data, and the discussion and conclusions are formed around spurious results.

There are lots of validity issues around subjective data, questionnaires and collecting data by observation. When answering questions, do respondents tell you what they think you want to know, do they try to show themselves off in a good light, do they really answer honestly, do they feel restricted in what they want to say by their employers? If you observe people to determine their behaviour, do they change their behaviour because they know they are being observed? You need to design your methodology to minimise the effects of some of these issues on the validity of your work.

At the end of your study you should self appraise the reliability and validity of your research holistically. It may be useful to consider both concepts on a scale of zero to 10. Zero could be anchored with a qualitative label of 'little value' or even 'worthless'. That is zero reliability and validity. Ten could be anchored with a qualitative label of 'first proof, beyond doubt, in a significant area of work'. Some intermediate labels may help in making judgements. How far can you expect to get on this hypothetical scale? In your dissertation you cannot be expected to score 10. Whilst your dissertation is a really significant piece of work for you, and whilst you will get some help, it is mostly limited to what you can do personally. You do not have unlimited time, unlimited money and a team of researchers to help. Do not score your work objectively on this zero to 10 scale; the scale is suggested as a way to think the issues through in your own mind.

Consider an objective to measure stress in site managers. That is, only to consider reliability and validity of the measurement of stress, not reliability and validity of the whole study. This objective may be founded in a problem summarised by Langford (1988) 'there is awareness amongst older site managers that retirement is all too often followed by a sudden death'. One part of the methodology might be to measure stress by observation of managers in the workplace, or by interviews, or by some kind of survey. The criteria used to make judgements about stress levels and/or the questions asked must be based on the literature. There are lots of studies and templates in the healthcare professions. You would be foolhardy to design your own measures of stress; that would constitute zero validity. If you design a measurement system adapted from the literature, this is a good foundation for asserting good reliability and validity. Perhaps a hypothetically respectable score, given limited resources, of five out of ten. However, better measures of stress are possible if site managers are subjected to a plethora of medical tests to measure heart rate, blood pressure, and cholesterol levels. There was a recent study where football managers were wired up to heart rate monitors during matches. The study tried to get a 10 out of 10 score for reliability and validity in measuring stress. Whilst it may be possible as part of university/industry funded research to get site managers wired up to heart rate monitors, you cannot reasonably do it as part of your dissertation. However, what you need to do is to research and write about what the authoritative sources suggest is the most valid way to measure stress; the gold standard methods using heart rate monitors, blood pressure measures, and cholesterol levels. Then, compare what you have been able to do, given your limited resources and access to site managers, with gold standard measurement methods.

Whilst you cannot be marked down for not 'scoring' 10 out of 10, you will get marks for reaching as high up the scale as you reasonably can. Also, you will get more marks for recognising where on the scale your work is. This would mean that you have good insight into the concepts of reliability and validity and that you understand the limits of what you have done. You will lose marks if you do not understand the limits of the reliability and validity of your work. You may be able to assert that some parts of your study are highly valid, but recognise that other parts are less so. You may start the process of recognising where your work is not as valid as you would like in your methodology chapter. You may expand on that in the conclusion chapter, with a sub-section titled 'limitations and criticisms of the study'.

Chapter 8 illustrates examples of how to use correlation coefficients to measure reliability and validity.

## 5.7 Analysis, results and findings

The methodology chapter provides a lead-in to the analysis, and the subsequent discussion and conclusions chapters. The terms, 'results, findings, and conclusions' are often used interchangeably in the literature and the media. It may be appropriate to consider the definitions in the context of your study. You need to be clear in your own mind how you apply these terms to your dissertation. The way you use the terms may be driven by whether your analysis is quantitative or qualitative. For quantitative work the terms 'analysis, results and findings' may be presented in the same chapter and in that order. They are key elements in helping you to meet study objectives; they are all focused on the objectives.

The analytical process is one whereby all raw data are brought together and sorted systematically. This may involve some numerical coding of qualitative labels. It may involve summarising the data in spreadsheets. Data, that is available numerically, can be subjected to a variety of academic analytical tools. It may be put into simple or complex formulae. There may be some adding, multiplication, calculations of means, medians, modes or standard deviations. There may be some comparison tables or simple frequency counts. More complex calculations may be performed using computer software. The strength of relationships between variables and significance levels may be determined. 'U' value or sound insulation calculations may be executed. If you have few calculations being performed 'by hand', these may be written out in full in this chapter. If there are many calculations, they may be best placed in an appendix. The raw data goes into the analytical process unstructured; it comes out structured.

After the analysis comes the results. Calculations terminate with an answer. The answer is the result. It may be a figure, such as 2.0 w/m$^2$, °C, or 28 N/mm$^2$, or 50%, or £50.00/m$^2$, or $p \leq 0.05$. As a figure, it is short, it is clear, it requires few words. If you are testing hypotheses, you should state the null hypothesis in full before each result. A contingency table used in chi-square calculations (see Chapter 8) may summarise frequency counts, or scatter diagrams can be used to show relationships between variables. Test results from laboratory work may be illustrated in graphs or line diagrams. If there are many results, they are likely to be presented in tables.

After results, there can be statements of findings that stem from results. If your quantitative study has only one variable, the finding mirrors the result, e.g. the study finds that: the 'U' value of this composite structure is 2.0 w/m$^2$, °C, or this concrete cube failed at 28 N/mm$^2$, or the mean quality score was 50%, or the cost of floor finishes was £50.00/m$^2$. If your quantitative study has two variables and you have tested a hypothesis, you should state as part of your findings that you either 'reject the null hypothesis' or 'cannot reject the null hypothesis'. You may also be able to assert findings such as the IV influences the DV, or method or group 'A' is better or lower cost than 'B'. Alternatively, the IV is not found to influence the DV, or method or group 'A' is not found to be better or lower cost than 'B'

The analysis, results and findings chapter will also summarise any demographic data that you may have collected. This may include ages, qualifications, and job positions of people involved in surveys. It may be most simply summarised in tables, rather than too many pie charts or histograms. If your work involves lots of analysis, it may be useful to summarise the results in a table towards the end of the chapter.

In qualitative studies, the raw data comprises words, possibly verbatim transcripts of interviews. The analytical process comprises ordering, labelling and sorting paragraphs of text. Transcripts and labelling are likely to be in appendices. There may be some frequency counts of key words and tables to integrate with the literature. A narrative will be derived from the tables. The tables and the narrative may be in the dissertation's main body. The boundary between analysis and results, and results and findings are not so clear. Labelling, sorting and frequency counting are part of the analysis. The process of writing the narrative may be thought of as part of the analytical process, but it is also the outcome (the result) and it asserts what is found.

## Summary of this chapter

You should design your study carefully. In your methodology chapter describe in detail what you have done so that it can be replicated by others, and justify why you have done it that way. Seven approaches to collecting data are identified. You should be proactive and innovative in collecting data; only use questionnaires if they are the best way to meet your objectives. Primary data sets are generated by the reseacher and secondary data sets are existing, compiled by others. Arguably, students use secondary data too infrequently. Objective data sets are facts uncoloured by people's opinions, whilst subjective data sets are people's views. For some studies, only subjective data are appropriate. For other studies, objective or hard data sets may be preferred, but if it is too difficult to secure that data, alternatively subjective measurements may be used. In such cases you should recognise the limitations this has on your study. If you use questionnaires, a basket of questions or multiple item scales should be used to measure variables. Only use your questionnaire to measure your variables, not to ask other interesting questions. If possible, use a questionnaire from an authoritative source, adapt if necessary to your study, and pilot it. Score and code responses so that they can be summarised in a spreadsheet. Invite respondents to add their comments, and write those comments up verbatim in an appendix. There are many analytical tools that can be used to analyse data that draw on the qualitative and quantitative techniques. If you demonstrate understanding of reliability and validity, and incorporate these concepts into your study, this gives you the possibility of achieving high marks. Reliability is consistency over time, and validity is whether a study achieves what it purports to achieve. Analysis is followed by results, which is followed by findings. Be careful to distinguish between the three and provide all of them in a chapter under this or a similar heading.

# 6 Qualitative data analysis

The objectives of each section of this chapter are:

6.1 Introduction and the process: to provide context for collecting qualitative data and to describe the process
6.2 Steps in the analytical process: to provide a template

## 6.1 Introduction and the process

Interviews are often the data collection technique used in qualitative data analysis. However, qualitative tools can also be applied to data collected in other ways, such as case studies, focus groups, reviewing transcripts of speeches, judgements in construction law cases, or other literature in all its forms. This section assumes the use of interviews, but the analytical process of ordering and coding is applicable to data collected by other approaches.

The qualitative goal is to define categories during the process of the study; the application of social science to the study and improvement of contemporary life depends upon the intimate understanding of respondents (McCracken 1988). Qualitative research looks for patterns of relationships between categories, whereas quantitative research seeks to delineate between categories; applying this to the analogy of the spider's web that was used in Chapter 2 to explain development of theory, the qualitative work seeks to look across an area of the web mesh, whilst the quantitative work looks at only one strand in the web. In qualitative work respondents may have difficulty in giving an answer; they may labour to articulate responses and the data you capture are rich. Open questions are the norm. Qualitative data aim to get insight into how respondents see and view the world. You may seek to gain access to cultural values held by respondents in a certain category, e.g. in a construction context, this may be craftspeople. There are lots of differences in life that need to be teased out, such that diverse societies can have appreciation of others. Again in construction, this may be younger workers seeing the world differently to older workers, or craftspeople differently to site managers, or

*Writing a Built Environment Dissertation: Practical Guidance and Examples.* Peter Farrell.
©2011 Peter Farrell. Published 2011 by Blackwell Publishing Ltd.

contractors differently to clients, or operational managers differently to strategic managers. There is the danger that you will go into a quantitative study with blinkers on; the problem is only as you see it. If you undertake some qualitative work speaking to others, even if it is only modest, it should help you take those blinkers off.

What is discovered in qualitative work cannot be assumed to exist widely. It does not seek to survey the terrain. You will *not* make the inference that what you have found in your very small sample exists in the population. Only quantitative methods, with large sample size, can make inferences. Interviews are used as a substitute for observation: they have limited validity. Interviewees may tell you what they think you want to know, or may speak to show themselves off in the best light. If you want an insight into their practice, observation may be better. However, observation may not be attainable; it is potentially time consuming and you may not get permission to do it, other than perhaps if you are a part-time student undertaking action-based research.

If you are careful in selecting potential people for interviews, response rates to requests should be in the region of 70% or 80%. Always follow the ethical code in your university; provide respondents with an information sheet in advance of your interview so that they may have time to digest its contents. Prepare carefully and do not offend respondents or breach confidentiality or anonymity. Do not be too ambitious in the level of seniority of people you ask. Interviews do not have to be face-to-face. It is quite reasonable that telephone or IT methods are employed, especially if you are only using the data to support quantitative work. A long telephone call is less resource intensive than travelling long distances, though you should pre-book the call with respondents. Interviews may be exploratory or may be used as the only method of collecting data. Students often ask 'how many people should I interview for in-depth qualitative work'? There is no definitive answer; it depends on the context of your objective, the precise type of analysis you propose, and whether interviews are your only method of data collection. If it is the only method, perhaps eight may be a guide figure for you to start thinking around. You may justify more or less; since this is an important element of your study, speak to your tutor.

Interviews may be considered on the continuum of unstructured, semi-structured and structured. In unstructured interviews you should have some pre-prepared themes that you wish to explore, but the precise nature of questions may be adapted to suit the individual interviewee and the answers you receive. Semi-structured interviews may have a list of definitive questions, but with some pre-prepared possible probe questions and the licence to probe still further and ask for more detail if required. Probing is a technique used to get more information when the response is unclear or incomplete. Probes include gestures such as nodding or neutral questions such as 'can you tell me more?' The answers you get may not answer the question; not that respondents may evade questions as sometimes happens with politicians, but possibly they misunderstand them. Structured interviews could almost be conducted as though a postal or electronic survey. There are no opportunities for follow-up questions or probing, although clarification may be given if the interviewee does not understand something. Structured interviews are moving away from the qualitative domain to the quantitative. Answers to questions may be limited to a number of closed responses that will be converted into quantitative data. Since the proposed analysis will be quantitative, the number of people interviewed may be higher. The reason for using such interviews is to improve response rates. For example, if your objective were to determine the attractiveness of energy-saving technologies to potential home buyers, you may target a sample by asking permission of estate agents to conduct short, structured interviews with members of the public who are browsing sales stands.

To do qualitative work in interviews properly, you do need to use a tape recorder. You can only do this with the permission of the interviewee. Try and pre-advise that you wish to tape record; some interviewees may be reluctant. If they are reluctant, you must respect their wishes. You will no doubt still get valuable data by just taking notes. The disadvantage of mere note taking is that in the context that you are trying to collect rich data, you want to be sure that you secure every single word, including intonation. Also, you may be less relaxed, it may take your concentration away from thinking about potential probing questions, and you may not able to observe facial expressions that may be 'telling' more than the words. If you do only take notes, it is important to write them up in full, ideally on the same day whilst the interview is still fresh in your mind. If you have been able to tape record, you can also make some brief notes during interviews as memory jolts. Type up interviews verbatim (word for word). Include your own observations at relevant parts; these may be about facial expressions, or opinions expressed with passion. Your transcripts should include all colloquialisms, slang, and anecdotes, but with any bad language dashed out. Be mindful that for every 1 hour of interview, the writing-up process can be as much as 5 hours, unless you have used specialist software support. Imagine the difficulty in playing the tape for a few seconds, then stopping the tape whilst you write, then re-winding to re-play to ensure your writing is accurate. There is no reason why you should not obtain some support from friends, family or elsewhere to undertake this task, providing, that is, you acknowledge the help you have had in the preliminary pages of your document, and that you personally check that the transcripts have been correctly transcribed. Place transcripts in an appendix, and do not name people interviewed, or give any indication of their employer. Person 'A' may be, for example, identified as an architect working for a medium-sized design practice in the south-east of England.

Your interviews may normally be of 30 minutes' duration, though longer may sometimes be justified. You need to respect the time of the interviewees, who are no doubt busy people. Try to be persuasive and informative at the introduction to the interview and then to create smooth conversational flow. At the outset, you will need to set the context for the interview, and explain what you hope to achieve from it. You may wish to ask some personal questions about a respondent's background, but do not ask for data that may be unnecessarily intrusive.

As part of your interview preparation, be clear about its 'purpose'. It could be: (a) at the early stage of your research, to help you explore or define the problem to be investigated, (b) at the early stage, before or after your theory and literature review, to help you set the study objective and define variables, (c) as part of a piloting process, (d) as part of data collection for the main part of your study, or (e) at the closing part, to help you develop discussion and conclusions. Plan the interview structure and questions. Be clear about the generic theme (or objective or variables) that you wish to explore, and about whether you wish to tease out the breadth or depth of a problem area. Ensure that all your questions are directed towards the theme; there should be no superfluous questions. Try to ask questions that will generate long answers, rather than one-line or even single word answers. If you do receive short answers, probe further. Prepare some potential probe questions. Number each question Q1, Q2, Q3 etc. Number any planned probe questions, e.g. two probe questions to question 1 as Q1a and Q1b. In the interview try to relax, but be alert in your thinking.

## 6.2 Steps in the analytical process

The analytical process is to take the raw data at the beginning and convert it, using a step-by-step process, into a final narrative. The narrative summarises, in a prosaic form, the literature and the raw data in preparation for conclusions and recommendations. There is the potential to include a separate discussion chapter between the narrative and the conclusions. However, since one role of a discussion is to integrate study findings with literature, and since the literature has already been integrated, there is the possibility to exclude a chapter specifically titled 'discussion'. If you decide not to write a discussion chapter, you should ensure that your narrative has a discussion element to it. The analytical process can be undertaken manually by photocopying, then cut and shuffle of pieces of paper. At the other end of the spectrum, sophisticated computer software can be used, such as QSR International XSight and NVivo7. If you are able to master use of software such as this, you should be given credit for it in your mark; but do not let your use of the software overtake the need for you to immerse yourself in the data and develop a narrative and conclusions with insight. The following assumes an intermediate position between cut and shuffle and specialist software; that is, using word processing software, such as Microsoft Word.

You will go through 10 steps and have 10 electronic files. Save all files carefully as master files, and when you revise them as work in progress, you may wish to resave with a file name that includes the current date. The process is one of sorting and condensing the data into a table at penultimate step 9, and then finally opening up the data from that table into a flowing narrative at step 10. Four of the 10 files contain the interview transcripts. There are two steps that add data to transcripts. Firstly, paragraphs of text are 'coded' with numbers and letters so that when data transcripts appear in their fourth version (step 8 below), you are able to trace their origins back to the source (step 3 below). Secondly, paragraphs of text are 'coded' with labels to allow later sorting of data within transcripts into like themes (step 7). The suggested steps and files are:

(1)   The research questions and objectives.
(2)   Interview questions and prompts.
(3)   Verbatim transcripts of interviews first copy; separate files for each interview, saved as file 3a, 3b, etc.
(4)   Verbatim transcripts of interviews second copy; originating from file 3.
(5)   Content analysis; frequency counts.
(6)   A basket of labels.
(7)   Verbatim transcripts of interviews third copy; originating from file 4.
(8)   Verbatim transcripts of interviews fourth copy; originating from file 7, paragraphs cut and paste (cut and shuffle) so that like labelled paragraphs appear together.
(9)   A new file, comprising tables.
(10)  The final narrative.

The early design work is key. Some steps are extremely time consuming, such as typing up interview transcripts. Some steps involve mere cut and paste computer work. The key steps where you immerse yourself in the data, and do the analysis, are steps 6, 7, 9 and 10.

The following detailed description is aligned to the example in Appendix D. It originates from work by Stott (2010). The key study objective is to determine whether the propensity to put completions before quality influences profit within the private housebuilding sector. It is based on interviews with ten housing site managers. The size of documents in such interviews can be large. For brevity in Appendix D, the following adjustments are made: (a) only transcripts of interviews with two site managers are included in file 3, (b) file 4, which comprises the second version of transcripts of interviews is not included, (c) file 7, the third version of transcripts is for one site manager only, and (d) file 8, which contains the fourth verbatim transcripts of interviews sorted under 17 labels, illustrates the data set for just one label of the 17.

Following the suggested steps above, the details are:

(1)   The research questions and objectives. The purpose of this file is to ensure that the questions you ask in interviews do not deviate from the origin of your study as described in the articulation of the problem, and the research questions and objectives derived therefrom. Save this with a file name '1'.

(2)   Interview questions and prompts. The purpose of this file is to record the specific interview questions that are designed to tease out answers to the generic research questions and objectives. These questions will include some general introductory questions, and to ensure that you comply with your university ethical code should include confirmation that details in the information sheet are clearly understood. Code each question Q1, Q2, etc. Code question prompts as Q1a, Q1b, etc. Save this with a file name '2'.

(3)   Verbatim transcripts of interviews first copy. The purpose of these files is to retain verbatim transcripts of interviews, and to start coding. Type-up each transcript. Ensure individual transcript has a heading to identify the interviewee with letters (person A, person B, etc).

Code each answer with numbers/letters, e.g. A-Ans 1, A- Ans1a, A-Ans1b, and for person B code each answer B-Ans 1, B- Ans1a, B-Ans1b, etc. This will allow the original source of the data to be traced at later stages. There is no necessity to code the questions such that they can be traced to the original source, since we will later delete the questions. If you ask probe questions spontaneously in the interview, and thus these questions will not be asked to every interviewee, to distinguish them from pre-prepared probes, start marking them with codes in the second half of the alphabet, e.g. Q1n. Add your own observations from notes you made in the interview; identify these notes as interviewer's observations, e.g. IntOb1. Numbers for these observations can run numerically across all interviewees. Since there will be several transcripts, save these with file names '3a', '3b', etc.

(4)   Verbatim transcripts of interviews second copy; originating from file 3. The 'purpose' of this file is to merge all the raw data from file 3 into one file. Also you may choose to delete non-essential data to remove some of the 'clutter'. Copy and paste files 3a, 3b, etc. into one file 4, keeping them all in order of person A, person B, etc. Optionally, delete (a) headings that have been used to identify interviewees (person A, person B, etc.), (b) all questions, and (c) all question prompts. If you do not delete these data sets, they will be retained until step 8. Save this with a file name '4'.

(5)    Content analysis; frequency counts. The purpose of this file is to record frequency counts. The simplest form of analysing qualitative data is content analysis; a word count of regularly used 'key words or phrases'. With file 4 open, using the 'find' function in the word processor, type in a key word or phrase that you observe occurs frequently. Count and record the number of hits you receive for each; ignore any hits in questions. You may decide that this will be the limit of your qualitative analysis. If this is the case, proceed to your narrative write-up in step 10, flavouring it with the number of hits you make. For example, transcripts of interviews with site managers may get high frequency hits against the word 'safety', and few hits against the word 'money'; the narrative write-up will reflect this balance. Alternatively, this content analysis will be the first part of more detailed analysis in steps 6 to 9.

The frequency counts of some key words for the two site managers are:

| | |
|---|---|
| Customer/s | 33 |
| Subs/subbies/subcontractors | 25 |
| Quality | 24 |
| Busy | 13 |
| Defect/s | 12 |
| Snag/s | 12 |
| Money | 8 |
| Rush | 7 |
| Weekend | 7 |
| Safety | 5 |
| Building inspector | 3 |
| Complaints | 2 |

(6)    A basket of labels; as the example below. The purpose of this file is to design and log your labels. These labels will not necessarily be the same words that are used for the content analysis. As an analyst, take a careful overview of the transcripts; what are the themes that are arising? Compile the basket (or qualitative codes); our example has seventeen. The labels may be single or perhaps two or three words long. These labels may be words or phrases used by interviewees or may come from the literature. As part of the process of condensing the data, arrange the labels that have the same generic theme into groups, and give each group a generic name. Our example shows that three labels 'culture', 'hierarchy' and 'specific periods' are grouped together into one group named 'propensity'. Give an abbreviation to each label, so that in later search and find functions in Word, it does not generate any hits with the interview transcripts, e.g. PROP1, PROP2, etc. Type-up the basket of labels, and include it as file 6. For the example in Appendix D, the labels and their abbreviations, sorted into generic group names, are:

Propensity

| | |
|---|---|
| PROP1 | Culture |
| PROP2 | Hierarchy |
| PROP3 | Specific periods |

Completions

| | |
|---|---|
| COMP1 | Workload |
| COMP2 | Resources |
| COMP3 | Supply chain |
| COMP4 | Communication |

Quality

| | |
|---|---|
| QUAL1 | Defects |
| QUAL2 | Quality control |
| QUAL3 | Workforce |

Profit

| | |
|---|---|
| PROF1 | Budget targets |
| PROF2 | Repeat business |
| PROF3 | Bonus payments |

Time

| | |
|---|---|
| T1 | Build programmes |
| T2 | Working conditions |
| T3 | Health and safety |
| T4 | Sales issues |

(7) Verbatim transcripts of interviews third copy; originating from file 4. The purpose of this file is to rearrange the data set into paragraphs of manageable length and label paragraphs in preparation for further rearrangement in step 8.

Break down or build-up the transcripts up into 'manageable paragraphs' of perhaps between two and five sentences long. Place a paragraph break where there may be a change in direction by the interviewee. If, in file 1, there are short paragraphs that originally resulted from short answers to main questions, you may merge those answers with answers to the probes that you should have followed-up with. It may be that one answer is a 'manageable paragraph'. Hopefully, some answers may be longer answers, and you will break them into manageable paragraphs.

Ensure the beginning of each paragraph is marked with a code, e.g. Ans1, Ans2a, Ans 3n. If there has been merging of answers, e.g. Ans1 and Ans1a are merged, cut out the Ans1a from the middle of the new paragraph and mark it at the beginning as 'Ans1 and Ans1a'.

Carefully read each paragraph as an analyst. With insight, allocate each paragraph a 'label' from your basket; what is the theme that this person is driving at that can be summed-up by the theme label? Add new labels to your basket if you think it appropriate. Type in the abbreviation for the label at the end of the paragraph, perhaps in bold text, to distinguish it from the main text. If you feel a paragraph addresses two or more themes, allocate it two or more labels. Put paragraph brackets around the text to which the label is relevant. Some text may be 'superfluous'; that is, transcript from interviews that deviated from study objectives. Do not code this type of data, and then it will not be taken forward into step 8. Do not put brackets around such superfluous data; this will help to distinguish superfluous from labelled data when you start file 8.

(8)   Verbatim transcripts of interviews fourth copy; originating from file 7. The purpose of this file is to rearrange the data set electronically by 'cut and paste' or 'cut and shuffle', so that like-labelled paragraphs appear together in new file 8. The transcripts for the two site managers in file 7 have 129 labelled paragraphs. Two labels are ascribed to 26 of these paragraphs. These 129 paragraphs will represent themselves in file 8 as 155 paragraphs ($129 + 26 = 155$).

Open up a new blank file, file 8. Resave file 7 as file 7a, with the intention of cutting paragraphs of data out of file 7a and rearranging them in file 8. At the end of the process file 7a will be a blank file except for 'superfluous' data. Have both files 7a and 8 open at the same time. In file 8, set up a series of separate pages that are blank, and type on each page headings that are each of your basket labels. In our example we have seventeen. With file 7a open, take each paragraph in order and cut and paste it to the appropriately headed page in file 8. Keep resaving both files 7a and 8. If your manageable paragraph has two labels (or more; our example does not have paragraphs labelled more than twice), *copy* and paste the paragraph twice (or more), into file 8, e.g. you may have coded one paragraph with two separate labels 'PROP1' and 'COMP1'. Continue with this process, until file 7a is blank except for superfluous data, and file 8 has 155 paragraphs arranged under seventeen basket labels. Recognise that file 8 is made up of paragraphs bolted together in a clumsy fashion that have no prosaic fit between each other. Delete all the labels at the end of each paragraph.

(9)   A new file, comprising tables. The purpose of this file is to bring to bring together in summary format the data from file 8 and the literature. There should be eight columns, and 17 rows, as partially illustrated in Table 6.1 and fully illustrated in Appendix D (file 9).

To complete data in the column 'content analysis', go back to file 7. Using the search and find function in Word, undertake a frequency count for each label. For our example, gives counts as follows: PROP1, 11; PROP2, 15; PROP3, 6; COMP1, 16; COMP2, 4; COMP3, 1; and COMP4, 2.

Proceed to complete remaining cells in the table, with insight. For the column literature source, go back to your review of the theory and the literature; what authoritative sources or studies address the theme labels that you have identified?

In our example, there is separate identifiable literature for each of the three labels 'culture', 'hierarchy' and 'specific period'. Therefore, in the literature column, the three rows are kept separate. However, in the group named 'completions', there is no separate identifiable literature for the four labels 'workload', 'resources', 'supply chain', and communication', so therefore the four rows are merged into one cell.

With insight, read file 8, and read the relevant parts of the literature. Labour at length over what the literature and your interviewees are really saying. Make your observation, or proposition or explanation (based on the evidence, using your skill, but not giving your personal opinion). Identify where there are inconsistencies within this complete data set (your respondents and the literature). The

**Table 6.1** Example of table for file 9 (note: the full exemplar table is in appendix D, file 9).

| Label number | Group name | Label qualitative | Content analysis | Literature source | Observation, proposition, explanation | Inconsistencies | Similarities |
|---|---|---|---|---|---|---|---|
| 1 | Propensity | Culture | 11 | Smith (2009) | | | |
| 2 | | Hierarchy | 15 | Brown (2005), Jones (1990) | | | |
| 3 | | Specific period | 6 | Constructing Excellence (2007) | | | |
| 4 | Completions | Workload | 16 | CIOB (2000), CIOB (2005) | | | |
| 5 | | Resources | 4 | | | | |
| 6 | | Supply chain | 1 | | | | |
| 7 | | Communication | 2 | | | | |
| Etc. to 17 | | | | | | | |

inconsistencies may be within the literature (you should have already written about these in the literature review), between the interviewees or between the literature and the interviewees. In the same way also identify similarities. Insert statements in the table. You may arrange lines in your table that allow you in some cases to give statements against individual labels, or to group similar themed labels together for generic statements.

(10)   The final narrative. The purpose of this part is to take the condensed data in the tables in file 9, and open up such data into a free-flowing narrative. File 9 is your plan from which you will write the narrative. With insight, read your table. Use the table as a framework to write-up a final narrative whole, in a prosaic form. Your narrative should flow, guided by each of the group titles and the labels within each. Your narrative will weave the inconsistencies and similarities together and, as the author, you will state your observations or propositions or explanations. It will bring together in summary form, with the literature, all the important things your interviewees told you. Your narrative may have many caveats and often used 'if' and 'maybe'. Consider whether the narrative should embrace a discussion, or whether a discussion will be a separate chapter. Ensure the narrative is a platform for the conclusion and recommendations. Recommendations may include a proposal for a quantitative study stemming from your qualitative work. Readers of your document should be very interested to read this narrative in detail, as it will be one of the most interesting and informative parts of your dissertation.

Finally, you will need to make a judgement about where to place these files in your dissertation, or indeed if to include them at all. The following suggestions are made:

(1)    The research questions and objectives. These will be in the introduction chapter anyway, and be repeated throughout the document. There is no need to include them again in an appendix.
(2)    Interview questions and prompts. Include as an appendix.
(3)    Verbatim transcripts of interviews first copy. Since file 4 will be placed in an appendix, there is no need to include file 3.
(4)    Verbatim transcripts of interviews second copy. Include as an appendix.
(5)    Content analysis; frequency counts. Include in the main body of the dissertation, in the analysis results and findings chapter.
(6)    The basket of labels. Include in the main body or appendix.
(7)    Verbatim transcripts of interviews third copy. Since file 8 will be placed in an appendix, there is no need to include file 7.
(8)    Verbatim transcripts of interviews fourth copy. Include as an appendix.
(9)    A new file, comprising tables. Include in the main body of the dissertation, in the analysis results and findings chapter.
(10)   The final narrative. Include in the main body of the dissertation, in the analysis results and findings chapter, but perhaps as part of the discussion.

The above is suggesting two copies of the interview transcripts by files 4 and 8; these may be long documents. If the appendix becomes too bulky, the alternative position is to include a CD-ROM disc on the inside back cover.

## Summary of this chapter

The collection of qualitative data will attempt to give insights into the world of others, and provide rich data. The method of collecting data for analysis in the middle of your dissertation may be entirely qualitative. Interviews are just one method of collecting qualitative data; they involve few people, and inferences are not made that what is found may be found in the whole population. If you have undertaken a quantitative study, it is better if you support that with some qualitative work to support the development of the problem, or to provide insight for writing the discussion and conclusions. Be sure to follow your university ethical codes. The steps in analysing qualitative data involve having verbatim transcripts of raw data. These are coded and sorted into like themes, and then summarised in tabular format with the literature. Similarities and differences are noted, and observations made. The tables are used as your plan to write a narrative that brings together all the data you have captured, including the literature. It is the narrative that is the final output of the qualitative analytical process. It may also incorporate the discussion. Conclusions must follow; recommendations may include a fully developed proposal for a quantitative study that can be used to make inferences across the population.

# 7 Quantitative data analysis: descriptive statistics

The objectives of each section of this chapter are:

7.1 Introduction: to provide context for statistical data collection and analysis
7.2 Glossary of symbols: to list the symbols
7.3 Calculations done manually or using software: to encourage use of both methods
7.4 Descriptive statistics: to identify and explain appropriate ways to summarise data, with examples

## 7.1 Introduction

Some students may wish to avoid statistical analysis; avoiding some of the complexities in statistical work is perfectly acceptable. It is certainly better to avoid complexities, rather than to undertake analysis where there is weak understanding of underpinning principles; misunderstandings are 'dangerous'. Therefore, know your limits. In business, manipulation of statistical data is pervasive; it is really difficult to avoid it completely. At a basic level, comparisons of salaries or costs require the analysis of numerical data. At a slightly more advanced level, there is a business need to understand cause and effect issues, and how samples taken from populations are used to predict behaviour or the buying inclination of people.

Statistical analysis dealing with one variable is known as descriptive statistics; inferential statistics are used to find links between two or more variables. Descriptive statistics are often sufficient to allow you to pass your dissertation. Their key use is to summarise data sets; a mass of numbers is condensed into easily digested and remembered figures. The use of descriptive statistics lends itself best to data that are measured at the categorical or better still, the interval level. Descriptive work with categorical data is limited, but if those are the data you have, you should still proceed with as much descriptive analysis as you can. Charts provide easily digested, visual representation of summarised data sets. The most popular charts are frequency histograms, line diagrams, pie charts, and scatter diagrams. Be mindful that tables may more succinctly summarise data than too many charts with too much colour.

*Writing a Built Environment Dissertation: Practical Guidance and Examples.* Peter Farrell.
©2011 Peter Farrell. Published 2011 by Blackwell Publishing Ltd.

Better marks can be given if you demonstrate understanding of inferential statistics. This is justified because there is more complexity in inferential statistics than descriptive statistics, and your dissertation aims to test ability in complex data analysis. Better marks for inferential statistics are also justified in a business sense; since after all, life is about answering the question, 'if I change 'A', what will happen to 'B'?' An understanding of descriptive statistics is required before you move on to inferential tests; further, an understanding of the principles of normal distributions is required. If you do not feel able to grasp the concepts of inferential statistical analysis, this is fine; concentrate on presenting your descriptive work as best as you can.

Many of the tests are named after the statisticians who developed them: e.g. Karl Pearson, 1857–1936; Charles Spearman, 1863–1945; William Gosset (the '*t*' test) 1876–1937; Frank Wilcoxon, 1882–1965; Ronald Fischer, 1890–1962; Frank Yates, 1902–1994; Andrei Kolmogorov, 1903–1987; Henry Mann, 1905–2000; Maurice Kendall, 1907–1983; Milton Friedman, 1912–2006; Lee Cronbach, 1916–2001; William Kruskal, 1919–2005; Allen Wallis, 1912–1998, John Tukey, 1915–2000.

## 7.2 Glossary of symbols

If you undertake statistical analysis, you may use conventional mathematical signs as symbols. They may be letters from the Greek or English alphabet in italic font. On a preliminary page of the dissertation, you should provide a glossary. Table 7.1 shows

**Table 7.1** Common statistical symbols.

| Statistical use | Symbol | Comments |
| --- | --- | --- |
| A single score | X | |
| Number of scores in a population | N | Capital |
| Number of scores in a sample | n | Lower case |
| Mean of a population | $\mu$ | Greek mu |
| Mean of a sample | $\bar{X}$ | |
| Standard deviation of a population | $\sigma$ | Greek lower-case sigma |
| Standard deviation of a sample | SD | Capital |
| The sum of | $\sum$ | Greek capital sigma |
| Significance level | $\alpha$ | Greek alpha |
| Pearson's chi-square statistic | $\chi^2$ | Greek Chi |
| Mann–Whitney statistic | U | |
| Wilcoxon's statistic | T | |
| Related *t* test | t | |
| Spearman's rho | $\rho$ | Greek rho |
| Pearson's product moment correlation | r | |
| Probability value | p | |
| The null hypothesis | $H_0$ | |
| The alternative hypothesis | $H_1$ or $H_A$ | |
| Degrees of freedom | df | |
| Equal to or less than | $\leq$ | |
| Equal to or more than | $\geq$ | |
| Infinity | $\infty$ | |

commonly used symbols. $H_0$ and $H_1$ are commonly used to denote the null and alternative hypotheses respectively. If you have several IVs, DVs or subject variables, you may wish to distinguish between them by labels $IV_1$, $IV_2$, $DV_1$, $DV_2$, SV1, SV2, and without going too far, objectives as $O_1$, $O_2$, etc. These symbols are less common, but there is no reason why you should not use them.

## 7.3 Calculations done manually or by using software

Analysis can be done 'manually' or with the help of software. On the one hand, provided you understand the principles underpinning formulae, it is logical that you use software to perform calculations. However, you may wish to include some manual calculations to demonstrate that you can manipulate the formulae; there is the possibility then to confirm results using software. It is far easier if, as a minimum, raw data are presented in a spreadsheet. Also, if you have large data sets, calculations by hand can be tedious and almost impossible. Performing calculations by computer will enable you to experiment or poke around your data as part of post hoc analysis.

Specialist statistical software, such as SPSS (Statistical Package for the Social Sciences, recently re-launched as PASW 18, Predictive Analysis Software) or Minitab, are sophisticated spreadsheets. They are more powerful and will execute many more tests than software such as Microsoft Excel. They will also manipulate the data more easily. For example, if executing a chi-square test, SPSS will do this on the basis of using the raw data, as presented in columns 2 and 3 of Table 8.1. However, using Excel, you are required to manually calculate actual and expected frequencies, and represent the data as detailed in Tables 8.3 and 8.4, before executing the calculation. It is likely that you should be able to complete a good quality dissertation doing calculations either manually or in Excel. If you are attempting some more complex tests, such as factor analysis, Cronbach's alpha or Kendal's tau (there are many other tests too), you may wish to invest some extra time learning to use the specialist software. Provided you demonstrate insight and understanding of the tests executed, you should gain extra marks for being able to demonstrate ability to use more complex software. Instructions for using Excel to produce charts, and undertake descriptive and inferential statistical tests, are provided in Appendix E.

Table 7.2 illustrates the tests that are shown manually in this text, those that can be executed by Excel, and those which need specialist software. Eleven of the more popular descriptive statistics are shown; all are demonstrated by manual calculation, and by Excel. Eleven of the more popular inferential tests are shown; five are selected as examples for manual calculation. They are the four most often used non-parametric tests (Pearson's chi-square, Mann–Whitney, Wilcoxon and Spearman's rho); as an example of a parametric test, and since it cannot be calculated in Excel, the related '$t$' test is also demonstrated manually. Five inferential tests are demonstrated by Excel in Appendix E (Pearson's chi-square, Mann–Whitney, Wilcoxon, unrelated $t$ test and Pearson's product moment correlation). At undergraduate level, it is unlikely that you will be executing the other tests (Kruskal-Wallis, unrelated ANOVA, and the related ANOVA); if you do use them, you will need to use specialist software. Non-parametric tests involve a process of ranking; an example of how to rank data is also provided.

**Table 7.2** Tests demonstrated manually in this text, and by Microsoft Excel; tests which need specialist software.

| | Test no | Test name | Demonstrated by manual calculations in this text | Demonstrated by Excel in Appendix E | Possible to execute by SPSS or other |
|---|---|---|---|---|---|
| Descriptive tests | 1 | Count | Yes | Yes | Yes |
| | 2 | Sum | Yes | Yes | Yes |
| | 3 | Mean | Yes | Yes | Yes |
| | 4 | Median | Yes | Yes | Yes |
| | 5 | Mode | Yes | Yes | Yes |
| | 6 | Standard deviation | Yes | Yes | Yes |
| | 7 | Variance | Yes | Yes | Yes |
| | 8 | Maximum | Yes | Yes | Yes |
| | 9 | Minimum | Yes | Yes | Yes |
| | 10 | Range | Yes | Yes | Yes |
| | 11 | Confidence interval | Yes | Yes | Yes |
| Inferential tests | 1 | Pearson's chi-square | Yes | Yes | Yes |
| | 2 | Mann–Whitney | Yes | Yes | Yes |
| | 3 | Wilcoxon | Yes | Yes | Yes |
| | 4 | Friedman | No | No | Yes |
| | 5 | Kruskal-Wallis | No | No | Yes |
| | 6 | Unrelated '$t$' test | No | Yes | Yes |
| | 7 | Related '$t$' test | Yes | No | Yes |
| | 8 | Unrelated ANOVA | No | No | Yes |
| | 9 | Related ANOVA | No | No | Yes |
| | 10 | Spearman's rho | Yes | No | Yes |
| | 11 | Pearson's product moment | No | Yes | Yes |

## 7.4 Descriptive statistics

Hypothetical examples 1 and 2 are used to illustrate the application of descriptive statistics:

Example 1 in Table 7.3 uses forecast student examination results at UK universities; data such as these, with wide range (0–100), and high frequency counts, lends themselves readily to descriptive statistics being used as a tool to summarise the whole data set. Data are forecast for the entire population; that is one million examinations.

Example 2 in Table 7.4 is a questionnaire survey, which comprised eight questions (1a–1h), each with five possible responses coded 0–4. Together, the eight questions are a basket, measuring one theme. Since this is hypothetical, the data could be a measure of anything, perhaps client satisfaction, with a high score indicating high satisfaction. There are 30 respondents; minimum score of zero and maximum score of 32 for each person in column 9. Scores for each person are converted to a percentage scale, e.g. respondent 9, scored 16 out of $32 = 50\%$, in column 10.

**Table 7.3** Example 2: hypothetical data set of forecast student examination results at UK universities.

| Exam mark | No of students with this mark | Exam mark | No of students with this mark | Exam mark | No of students with this mark | Exam mark | No of students with this mark |
|---|---|---|---|---|---|---|---|
| 0 | | | | | | | |
| 1 | 0 | 26 | 200 | 51 | 31 300 | 76 | 7500 |
| 2 | 0 | 27 | 300 | 52 | 33 500 | 77 | 6100 |
| 3 | 0 | 28 | 300 | 53 | 36 000 | 78 | 5100 |
| 4 | 0 | 29 | 600 | 54 | 37 500 | 79 | 4200 |
| 5 | 0 | 30 | 700 | 55 | 39 000 | 80 | 3300 |
| 6 | 0 | 31 | 900 | 56 | 40 100 | 81 | 2600 |
| 7 | 0 | 32 | 1200 | 57 | 41 400 | 82 | 2100 |
| 8 | 0 | 33 | 1600 | 58 | 41 800 | 83 | 1600 |
| 9 | 0 | 34 | 2100 | 59 | 41 400 | 84 | 1200 |
| 10 | 0 | 35 | 2600 | 60 | 40 100 | 85 | 900 |
| 11 | 0 | 36 | 3300 | 61 | 39 000 | 86 | 700 |
| 12 | 0 | 37 | 4200 | 62 | 37 500 | 87 | 600 |
| 13 | 0 | 38 | 5100 | 63 | 36 000 | 88 | 300 |
| 14 | 0 | 39 | 6100 | 64 | 33 500 | 89 | 300 |
| 15 | 0 | 40 | 7500 | 65 | 31 300 | 90 | 200 |
| 16 | 0 | 41 | 9100 | 66 | 29 000 | 91 | 100 |
| 17 | 0 | 42 | 10 600 | 67 | 26 400 | 92 | 75 |
| 18 | 0 | 43 | 12 500 | 68 | 24 000 | 93 | 50 |
| 19 | 0 | 44 | 14 600 | 69 | 21 400 | 94 | 50 |
| 20 | 25 | 45 | 16 600 | 70 | 19 000 | 95 | 25 |
| 20 | 50 | 46 | 19 000 | 71 | 16 600 | 96 | 0 |
| 22 | 50 | 47 | 21 400 | 72 | 14 600 | 97 | 0 |
| 23 | 75 | 48 | 24 000 | 73 | 12 500 | 98 | 0 |
| 24 | 100 | 49 | 26 400 | 74 | 10 600 | 99 | 0 |
| 25 | 200 | 50 | 29 000 | 75 | 9100 | 100 | 0 |
| Total frequency counts | 500 | | 219 900 | | 742 600 | | 37 000 |

Total '$n$' = 500 + 219 900 + 742 600 + 37 000 = 1 000 000

Descriptives: count ($n$) = 1 000 000, sum = 58 000 000, mean = 58%, median = 58%, mode = 58%, standard deviation = 10%, variance = 100, maximum = 95%, minimum = 20%, range = 76%, confidence interval ±0.11%

Emphasis is given in this section to plotting frequency histograms. With a small sample size ($n$), if raw data sets are entered into a spreadsheet, they need to be re-arranged to allow frequency histograms to be plotted. For example, Excel will not plot a frequency histogram for the total score for each respondent, as shown in Table 7.4, column 10. Also, if each individual number on the 101 point scale was used for a histogram, the frequency counts would likely be very low, e.g. the highest frequency count is for the number 53, which occurs six times; the frequency count for most numbers is just one or two. To get higher frequency counts for each histogram bar, thus making the histogram meaningful, scores need to be re-arranged into convenient class intervals.

**Table 7.4** Example 2: hypothetical data set from a questionnaire survey, which comprising eight questions.

| | 1. Var1a | 2. Var1b | 3. Var1c | 4. Var1d | 5. Var1e | 6. Var1f | 7. Var1g | 8. Var1h | 9. Var1 total | 10. Var1 percentage |
|---|---|---|---|---|---|---|---|---|---|---|
| 1 | 0 | 2 | 2 | 2 | 1 | 1 | 3 | 2 | 13 | 40 |
| 2 | 1 | 1 | 2 | 0 | 1 | 3 | 3 | 3 | 14 | 44 |
| 3 | 3 | 2 | 3 | 2 | 1 | 3 | 1 | 2 | 17 | 53 |
| 4 | 3 | 1 | 2 | 1 | 2 | 2 | 2 | 2 | 15 | 47 |
| 5 | 2 | 3 | 4 | 4 | 2 | 2 | 1 | 2 | 20 | 63 |
| 6 | 0 | 2 | 1 | 2 | 3 | 3 | 3 | 4 | 18 | 56 |
| 7 | 2 | 1 | 2 | 2 | 3 | 2 | 4 | 2 | 18 | 56 |
| 8 | 3 | 2 | 0 | 2 | 2 | 1 | 1 | 0 | 11 | 34 |
| 9 | 3 | 1 | 2 | 2 | 2 | 2 | 1 | 3 | 16 | 50 |
| 10 | 0 | 2 | 2 | 3 | 3 | 1 | 2 | 2 | 15 | 47 |
| 11 | 3 | 1 | 4 | 0 | 3 | 0 | 3 | 3 | 17 | 53 |
| 12 | 4 | 4 | 0 | 3 | 3 | 3 | 1 | 1 | 19 | 59 |
| 13 | 3 | 2 | 2 | 2 | 3 | 2 | 3 | 2 | 19 | 59 |
| 14 | 0 | 4 | 4 | 2 | 3 | 0 | 3 | 4 | 20 | 63 |
| 15 | 3 | 4 | 0 | 4 | 2 | 2 | 1 | 4 | 20 | 63 |
| 16 | 3 | 1 | 2 | 0 | 2 | 2 | 3 | 2 | 15 | 47 |
| 17 | 1 | 3 | 3 | 3 | 3 | 2 | 3 | 4 | 22 | 69 |
| 18 | 3 | 2 | 0 | 2 | 1 | 0 | 3 | 4 | 15 | 47 |
| 19 | 0 | 2 | 2 | 2 | 2 | 2 | 1 | 1 | 12 | 38 |
| 20 | 3 | 1 | 3 | 2 | 2 | 2 | 3 | 1 | 17 | 53 |
| 21 | 3 | 3 | 3 | 2 | 4 | 3 | 4 | 0 | 22 | 69 |
| 22 | 4 | 3 | 4 | 2 | 3 | 3 | 3 | 1 | 23 | 72 |
| 23 | 3 | 4 | 3 | 3 | 3 | 4 | 2 | 2 | 24 | 75 |
| 24 | 3 | 2 | 3 | 2 | 2 | 2 | 1 | 4 | 19 | 59 |
| 25 | 0 | 4 | 2 | 4 | 1 | 3 | 3 | 1 | 18 | 56 |
| 26 | 4 | 1 | 3 | 2 | 2 | 2 | 1 | 2 | 17 | 53 |
| 27 | 4 | 3 | 2 | 2 | 2 | 2 | 1 | 1 | 17 | 53 |
| 28 | 1 | 4 | 2 | 1 | 2 | 1 | 3 | 2 | 16 | 50 |
| 29 | 2 | 2 | 2 | 1 | 2 | 1 | 3 | 2 | 15 | 47 |
| 30 | 2 | 2 | 2 | 1 | 3 | 1 | 3 | 3 | 17 | 53 |
| Count | 30 | 30 | 30 | 30 | 30 | 30 | 30 | 30 | 30 | 30 |
| Sum | 66 | 69 | 66 | 60 | 68 | 57 | 69 | 66 | 521 | 1628 |
| Mean | 2.2 | 2.3 | 2.2 | 2.00 | 2.27 | 1.90 | 2.30 | 2.20 | 17.37 | 54.27 |
| Median | 3 | 2 | 2 | 2 | 2 | 2 | 3 | 2 | 17 | 53 |
| Mode | 3 | 2 | 2 | 2 | 2 | 2 | 3 | 2 | 17 | 53 |
| Standard deviation | 1.37 | 1.09 | 1.16 | 1.05 | 0.78 | 0.99 | 1.02 | 1.19 | 3.11 | 9.79 |
| Variance | 1.89 | 1.18 | 1.34 | 1.10 | 0.62 | 0.99 | 1.04 | 1.41 | 9.69 | 95.79 |
| Maximum | 4 | 4 | 4 | 4 | 4 | 4 | 4 | 4 | 24 | 75 |
| Minimum | 0 | 1 | 0 | 0 | 1 | 0 | 1 | 0 | 11 | 34 |
| Range | 5 | 4 | 5 | 5 | 4 | 5 | 4 | 5 | 14 | 42 |
| Confidence interval | | | | | | | | | | 3.65 |

Therefore, three further columns of data are required. Table 7.4, column 10, is represented in Table 7.5, column 1. The first step is to arrange the scores into ascending order, as column 2, labelled var1 sort ascendancy. The next stage is to place the data into ranges. In this case the chosen range is 5 points' wide, 30–34, 35–39, etc. as shown in column 3, labelled range.

**Table 7.5** Example 2: data in column 10 of Table 7.4 represented and re-arranged into class interval. Frequency counts for hypothetical vars2 to 6.

| | 1. Var1 % score | 2. Var1sort ascendency | 3. Class interval | Frequency counts | | | | | |
| --- | --- | --- | --- | --- | --- | --- | --- | --- | --- |
| | | | | 4. Var1 | 5. Var2 | 6. Var3 | 7. Var4 | 8. Var5 | 9. Var6 |
| 1 | 40 | 34 | 30–34 | 1 | 1 | 4 | 0 | 5 | 4 |
| 2 | 44 | 38 | 35–39 | 1 | 1 | 3 | 1 | 4 | 2 |
| 3 | 53 | 40 | 40–44 | 2 | 1 | 2 | 2 | 3 | 4 |
| 4 | 47 | 44 | | | | | | | |
| 5 | 63 | 47 | 45–49 | 5 | 2 | 3 | 2 | 2 | 2 |
| 6 | 56 | 47 | | | | | | | |
| 7 | 56 | 47 | | | | | | | |
| 8 | 34 | 47 | | | | | | | |
| 9 | 50 | 47 | | | | | | | |
| 10 | 47 | 50 | 50–54 | 8 | 3 | 3 | 2 | 1 | 5 |
| 11 | 53 | 50 | | | | | | | |
| 12 | 59 | 53 | | | | | | | |
| 13 | 59 | 53 | | | | | | | |
| 14 | 63 | 53 | | | | | | | |
| 15 | 63 | 53 | | | | | | | |
| 16 | 47 | 53 | | | | | | | |
| 17 | 69 | 53 | | | | | | | |
| 18 | 47 | 56 | 55–59 | 6 | 4 | 4 | 3 | 0 | 1 |
| 19 | 38 | 56 | | | | | | | |
| 20 | 53 | 56 | | | | | | | |
| 21 | 69 | 59 | | | | | | | |
| 22 | 72 | 59 | | | | | | | |
| 23 | 75 | 59 | | | | | | | |
| 24 | 59 | 63 | 60–64 | 3 | 8 | 3 | 4 | 1 | 2 |
| 25 | 56 | 63 | | | | | | | |
| 26 | 53 | 63 | | | | | | | |
| 27 | 53 | 69 | 65–69 | 2 | 6 | 3 | 4 | 3 | 3 |
| 28 | 50 | 69 | | | | | | | |
| 29 | 47 | 72 | 70–74 | 1 | 3 | 2 | 5 | 4 | 2 |
| 30 | 53 | 75 | 75–79 | 1 | 1 | 3 | 7 | 7 | 5 |
| Count | | | | 30 | 30 | 30 | 30 | 30 | 30 |
| Mean | | | | 54.27 | 57.33 | 51.67 | 61.50 | 53.67 | 52.67 |
| SD | | | | 9.79 | 10.56 | 14.64 | 12.12 | 18.19 | 15.47 |

Column 4, labelled var1 frequency counts, provides the counts for each of the ranges. Appendix E gives the Excel instructions to produce a frequency histogram, using the data in columns 3 and 4. The frequency histogram is shown in Figure 7.1a. Columns 4–9 of Table 7.5, give further frequency counts for hypothetical vars2 to 6. These counts hypothetically originate from data such as those for var1. They are used to illustrate frequency histograms with appearances that do not resemble normal distributions; the histograms are in Figures 7.1b–1f.

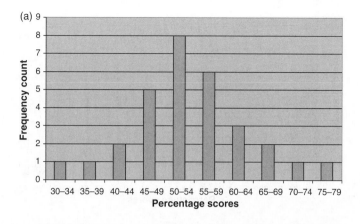

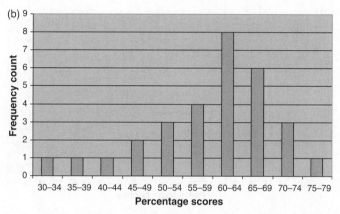

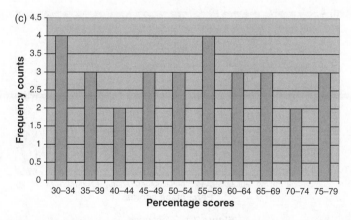

**Figure 7.1** Example 2: (a) Frequency histogram for var1, resembling a normal distribution. (b) Frequency histogram for var2, resembling a skewed distribution. (c) Frequency histogram for var3, resembling a flat distribution. (d) Frequency histogram for var4, resembling a stepped distribution. (e) Frequency histogram for var5, resembling a U-shaped distribution. (f) Frequency histogram for var6, resembling a random distribution, with no pattern.

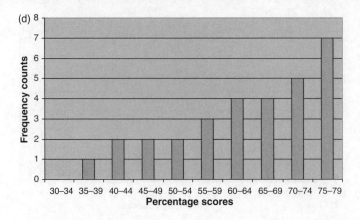

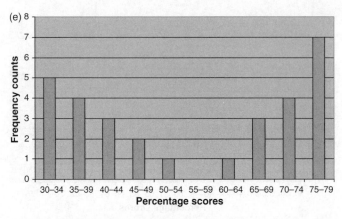

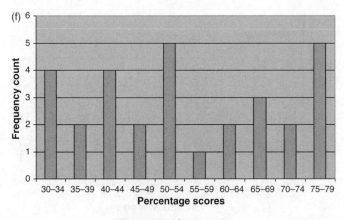

**Figure 7.1** (*Continued*).

**Table 7.6** Arbitrary numbers for ranking.

| | Var1 | | | Var2 | | |
| --- | --- | --- | --- | --- | --- | --- |
| | 1. Score | 2. Rank | 3. Score | 4. Score in ascending order | 5. Initial rank | 6. Mean ranks |
| 1 | 10 | 1 | 8 | 5 | 1 | 1 |
| 2 | 15 | 3 | 12 | 8 | 2 | 2 |
| 3 | 92 | 10 | 15 | 10 | 3 | 3.5 |
| 4 | 52 | 8 | 20 | 10 | 3 | 3.5 |
| 5 | 25 | 4 | 10 | 11 | 5 | 5 |
| 6 | 14 | 2 | 12 | 12 | 6 | 7 |
| 7 | 40 | 7 | 5 | 12 | 6 | 7 |
| 8 | 30 | 6 | 10 | 12 | 6 | 7 |
| 9 | 28 | 5 | 12 | 15 | 9 | 9 |
| 10 | 89 | 9 | 11 | 20 | 10 | 10 |

## Ranking

It may be necessary to rank data to produce frequency histograms; ranking is also used in non-parametric tests later in Chapter 8. Table 7.6 is a list of arbitrary numbers to be ranked. To facilitate ranking, particularly if you have a large data set, you may find it easiest to re-arrange the data in ascending order, before allocating a rank. The 'sort ascending' function in Excel will do this for you. Ranking can be lowest or highest number scoring 1 (ascending or descending). In the three non-parametric tests illustrated later, it is always the lowest number ranked 1. The one issue of complication is where scores are the same, and thus have joint ranks. It is dealt with in the same sort of way that golfers are positioned on a leader board. Joint ranks are averaged; if there is an even number of scores that are the same, a half point ranking arises. In Table 7.6, Var 1 (column 1) scores are all different; their ascending rank in column 2 is self-explanatory. Var 2 scores (column 3) are rearranged in ascending order in column 4, for simplicity. Column 5 indicates the ranking before mean ranks are determined. There are two instances of tied ranks. The two number 10s are jointly ranked 3, though taking up positions 3 and 4. Consequently, there can be no rank 4 in the initial ranking, and the next rank is position 5 (for the number 11). The mean of 3 and 4 is 3.5; the mean rank in column 6 of the two number 10s, is thus 3.5. The three number 12s, are ranked 6, though taking up positions 6, 7 and 8. Consequently, there can be no rank positions 7 or 8 in the initial ranking, and the next rank is 9 (for the number 15). The mean of 6, 7 and 8 is 7; the mean rank in column 6 of the three number 12s is thus 7.

## Normal distributions: measures of central tendency (mean, median and mode)

The bell-shaped normal distribution is represented in Figure 7.2 (source: Syque 2010). Normal distributions might be expected in the measurement of many variables to do with living organisms. The normal distribution represents frequency counts of measures across

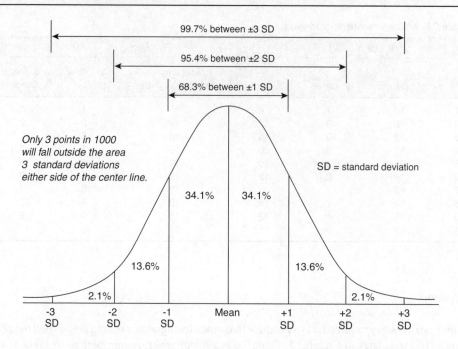

**Figure 7.2** Features of the normal distribution. Source: Syque, 2010.

whole populations; thus, frequency counts are very high, often multiple millions. Using example 1 and the data in Table 7.3 for forecast student examination results as an illustration, the features of a normal distribution are that:

The mean, median and mode are all the same value; in our example 58%
68.26% of scores should lie between ±1 standard deviation of the mean; in our example
  680 000 students should be expected to score between 48 and 68%
95.4% of scores should lie between ±2 standard deviations of the mean; in our example
  950 000 students should be expected to score between 38 and 58%

Mean, median and mode are known as measures of central tendency. The mean is computed by a simple formula, but more loosely, outside statistics, the term 'average' is often used. It is the sum of all scores divided by the number of scores. The median is the score that would be in the middle position, if all scores were arranged in ascending or descending order. The mode is the score that occurs most frequently, the highest bar on a frequency histogram. The data in Table 7.3 is used to calculate the mean, median and mode, thus:
  The mean is expressed by the formula:

$$\mu = \frac{\sum X}{N}$$

Where $\mu$ = population mean, $\sum$ = the sum of, X = each score and N = number of scores

The sum of all scores added together ($\sum X$) would be expected to be 58 000 000. Divided by 1 000 000, the expected mean score is 58%.

The median is the score that would be in the middle position, if the 1 000 000 individual scores were arranged in ascending or descending order; the score that is expected to appear in the 500 000th (half a million) position is 58%.

The mode, as the score that occurs most frequently, is 58% which has an expected frequency count of 41 800.

Not all data sets resemble a normal distribution. Therefore, mean median and modes do not always have the same score. They can all individually summarise a data set but, depending on the distribution of the numbers, one of them may be better than the other two. Eyeball observation of the frequency histogram is necessary to decide which is most appropriate. The mode is not used often; it is not likely to be a good measure of central tendency if the distribution of numbers differs radically from a normal distribution. Its weaknesses are that to determine the mode, numbers in other parts of the data set are not considered; also, there could be two different modes at different ends of a range of figures, e.g. the precise same highest frequency count at 40% and at 60%. The median is arguably better than mode as a measure of central tendency. If the data comprise an even number, at first glance there may be no median, e.g. 10 pieces of data, the fifth and sixth number, when arranged in order, are not the same. In that case, the median should be the average (mean) of the two numbers. However, the median also does not consider other numbers in the data. Therefore, the mean figure is used far more frequently as a measure of central tendency; it does consider the value of each number. To emphasise, this does not exclude the use of mode and median. The mean may not be appropriate if there is a small data set with one or two extreme or rogue values, e.g. a data set of 20, 20, 30, 40, 50, and 500 is not truly summarised by the mean, which is 660/6 = 110. It is appropriate to calculate all three, and then base subsequent findings and discussion on the most appropriate numbers, always keeping in mind that it should all be considered in the context of eyeball observation of the frequency histogram.

In Table 7.3, there are frequency counts for each measure of the variable that is possible; there are 101 points on the scale 0–100, and the sum of the frequency counts is 1 000 000. Frequency counts are calculated from the standard normal distribution table in Appendix F; the figures represent what is expected for a total $n$ count of one million. The frequency count of students expected to obtain the mean score of 58% is 41 800; the frequency count of students expected to obtain the score of 20% is 25. The theoretical distribution shows zero students scoring 0–19% and 96–100%; in practice, there are likely to be some scores in these bands.

The frequency histogram is shown in Figure 7.3. The vertical axis is frequency counts, reaching the maximum of 41 800 for the mean score of 58%. The horizontal axis is possible examination scores in the range 0–100%. In such histograms, the vertical scale is always frequency counts; the horizontal scale represents scores for the variable being measured/counted. If the variable being measured was annual earnings of construction workers, the horizontal axis would be £s with a 98% range of perhaps £10 000 to perhaps £200 000 per annum or higher; or if it were height of adult males, it would be millimetres, with a 98% range from say 1400–2100 mm.

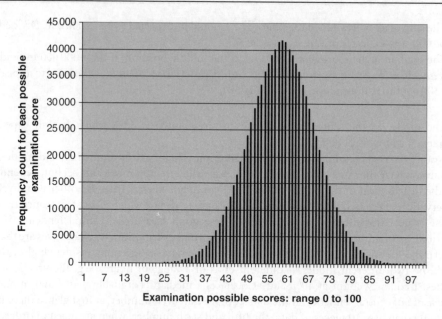

**Figure 7.3** Example 1: frequency histogram for forecast student examination results, $n = 1\,000\,000$.

By eyeball observation, it can be asserted that Figure 7.3 resembles a normal distribution. The difference between the normal distribution and the frequency histogram is that the frequency count for a true normal distribution is much higher. Therefore, in a normal distribution, instead of drawing each individual vertical bar in a histogram, just the curve is drawn, as illustrated in Figure 7.2, since there are many bars. This latter point is not so obvious if the horizontal scale is a percentage scale. There are only 101 points on the scale and expected scores occur only in the range 20–95%; there are only 76 bars. However, if the horizontal scale were £s, there would be more expected points on the horizontal scale, and therefore more vertical bars. Therefore, with more vertical bars and much higher $n$, the curve is most appropriate.

Chapter 8, section 8.4, illustrates two further methods, other than eyeball observation, to make judgements about whether your data set may be considered to resemble a normal distribution.

## Measures of spread: range, standard deviation, variance

Whether it is the mean, median or mode chosen as the lead measure of central tendency, on their own they cannot accurately summarise a data set. It may be possible that many of the numbers are clustered closely around the mean, with few at the extremities. Alternatively, the frequency count of numbers close to the mean may be low, and the count for numbers at the extremities is high. One of the frequency histogram shapes in figures 7.1 a-f, arising from example 2, may be similar to your data. Each shape gives potential for discussion in the dissertation. This discussion can be based on eyeball observation and a numerical value as a measure of spread. The simplest measure of spread is the range; the range for the data set in Table 7.3 is $95 - 20 + 1 = 76$. As a measure it is of limited value; as with median and mode, numbers in other parts of the data set are not considered. The most often used measure of spread, particularly if the data set is from a sample representing a population, is standard deviation. The unit of measure for the standard deviation is the unit being used to measure the variable on the horizontal scale, e.g. percentage score, or £s, or millimetres. The formula for calculating the standard deviation for a sample is:

$$\sqrt{\frac{\sum (X-\bar{X})^2}{n-1}}$$

When calculating the standard deviation of a whole population, the use of $-1$ does not apply. The formula uses all scores, and compares how far they all deviate from the mean. This is performed by the part of the formula, thus: $\sum (X-\bar{X})^2$. It is the longest part of the calculation, and is illustrated in Table 7.7. The data are used from Table 7.5 column 2; alternatively the figures in Table 7.4, column 10 could be used. The mean is deducted from each individual score, the difference is squared, and then the total for all differences is calculated as 2777.87.

Outside the table:

step 6: find the mean of all the differences squared, by dividing by $n - 1$

$$2777.87/30 - 1 = 2777.87/29 = 95.78$$

step 7: to convert the mean of all the differences squared to the mean of all the differences, take the square root

$$\sqrt{95.78} = 9.78$$

If you use Excel, note that it does use $-1$ when it computes the calculation. The standard deviation calculation using $n$ and not $n - 1$ is thus:

$$\sqrt{2777.87/30} = \sqrt{92.59} = 9.62$$

It is now appropriate to consider the standard deviations and frequency distributions for Figures 7.1a–f. Six possible distributions are illustrated for vars1 to 6: normal, skewed normal, flat, stepped, U-shaped and random. Each frequency histogram has a story to tell; potential weakness in the data set of the latter five may be explored. In the context of a questionnaire, a distribution close to normal may be anticipated. A skewed or stepped distribution may indicate a weakness in the width of the question scale. If it is skewed or stepped towards a high score, the high points on the questionnaire scale may not be appropriately tough, e.g. five possible question responses, with scores 0–4, and labelled poor, fair, satisfactory, good and very good may be better given as poor, satisfactory, good, very good and excellent; or better as a six-point scale, poor, fair, satisfactory, good, very good, and excellent. Flat, U-shaped or

**Table 7.7** Steps 1 to 5 for the standard deviation; data taken from example 2, table 7.5, column 2.

| | Step 1. Arrange the data in Var1sort ascendency X | Step 2. State the mean X̄ | Step 3. Column 1 minus column 2 (X − X̄) | Step 4. Column 3 squared (X − X̄)² |
|---|---|---|---|---|
| 1 | 34 | 54.27 | −20.27 | 410.87 |
| 2 | 38 | 54.27 | −16.27 | 264.71 |
| 3 | 40 | 54.27 | −14.27 | 203.63 |
| 4 | 44 | 54.27 | −10.27 | 105.47 |
| 5 | 47 | 54.27 | −7.27 | 52.85 |
| 6 | 47 | 54.27 | −7.27 | 52.85 |
| 7 | 47 | 54.27 | −7.27 | 52.85 |
| 8 | 47 | 54.27 | −7.27 | 52.85 |
| 9 | 47 | 54.27 | −7.27 | 52.85 |
| 10 | 50 | 54.27 | −4.27 | 18.23 |
| 11 | 50 | 54.27 | −4.27 | 18.23 |
| 12 | 53 | 54.27 | −1.27 | 1.61 |
| 13 | 53 | 54.27 | −1.27 | 1.61 |
| 14 | 53 | 54.27 | −1.27 | 1.61 |
| 15 | 53 | 54.27 | −1.27 | 1.61 |
| 16 | 53 | 54.27 | −1.27 | 1.61 |
| 17 | 53 | 54.27 | −1.27 | 1.61 |
| 18 | 56 | 54.27 | 1.73 | 2.99 |
| 19 | 56 | 54.27 | 1.73 | 2.99 |
| 20 | 56 | 54.27 | 1.73 | 2.99 |
| 21 | 59 | 54.27 | 4.73 | 22.37 |
| 22 | 59 | 54.27 | 4.73 | 22.37 |
| 23 | 59 | 54.27 | 4.73 | 22.37 |
| 24 | 63 | 54.27 | 8.73 | 76.21 |
| 25 | 63 | 54.27 | 8.73 | 76.21 |
| 26 | 63 | 54.27 | 8.73 | 76.21 |
| 27 | 69 | 54.27 | 14.73 | 216.97 |
| 28 | 69 | 54.27 | 14.73 | 216.97 |
| 29 | 72 | 54.27 | 17.73 | 314.35 |
| 30 | 75 | 54.27 | 20.73 | 429.73 |
| Step 5: $\sum(X − \bar{X})^2$ | | | | 2854.35 |

random distributions may indicate some significant structural faults in questionnaire design or sampling.

It can be seen that the standard deviation for Figure 7.1a (which approximates to a normal distribution) is 9.78%. But Figure 7.1e, which is not normal, has a higher frequency of scores further away from the mean, and thus a wider spread. The pictorial representation of the wider spread is summarised by one figure; a larger standard deviation of 18.19%. In your dissertation you may choose to present the frequency histograms and the standard deviations. In some circumstances, however (e.g. in conversation), it may not be appropriate to present histograms; the standard deviation statistic is therefore useful, since it is one figure that summarises the frequency histogram. Also, in findings and discussion, useful insight is possible by comparing standard deviations between data sets, or against known industry norms, e.g. standard deviation of 10% for student examination scores.

In normal distributions, we have stated that: (a) 68.26% of scores should lie between $\pm 1$ standard deviation of the mean, and (b) 95.4% of scores should lie between $\pm 2$ standard deviations of the mean. If the distribution is not normal, the 68.26% and 95.4% percentages do not apply; but irrespective of whether a frequency histogram resembles a normal distribution, the standard deviation calculation still gives one figure that can be used (alongside visual observation of the histogram, and a measure of central tendency, e.g. the mean), to summarise the data.

Step 6 in the standard deviation calculation, in our example the figure 95.78, gives a statistic known as the variance. Variance is sometimes computed by standard variation squared; again in our example $9.78^2 = 98.43$. The derivation of the formula shows it is in fact the other way round. Variance is a step before computation of standard deviation; standard deviation can be called the square root of the variance. Variance is therefore also a measure of spread of scores from the mean. However, since it is a high figure, it cannot be considered on the horizontal scale of the frequency histogram. In the context of our example, a score of 95.78 is not a useful eyeball measure in the context of the entire width of the horizontal scale of 0–100 or a range of 76. Because of this, standard deviation is more often used than variance.

In Table 7.3, tutors may reasonably expect a mean score of circa 58%, a standard deviation of 10%, and that the frequency histogram would replicate a normal distribution. These figures, by experience, may be benchmark standards. A mean score substantially more than 58% may indicate an examination that has been too easy, or if less than 58%, too difficult. A standard deviation of more than 10% may indicate a class where students who are good in a subject have been allowed to flourish, but less capable students have been left behind. A standard deviation of less than 10% may indicate a class where capable students have not been allowed to flourish.

## Standard score: the Z score

The area under a normal distribution curve can be considered equal to 1.0, and is linked into probabilities and percentages. The areas to the left and right of the central mean, median and mode position, each have an area of 0.50. Z scores allow you to take an individual score on a horizontal scale and calculate what percentage of scores lie to the left and right of that score on the distribution. This can be done, irrespective of whether the horizontal scale is examinations scores, annual earnings of construction workers or any other variable; and therefore scores, again irrespective of the variable on the horizontal scale, can all be standardised as z scores. Z scores below the mean are negative. Take the examination scores in Table 7.3. We have already calculated the mean as 58% and the standard deviation as 10%. We have stated that a feature of a normal distribution is that 68.26% of scores should lie between $\pm 1$ standard deviation of the mean. Therefore, an examination score of 48% should have 15.87% of scores less than this, and an examination score of 68% should have 15.87% of scores more than this. To prove this, the procedure is first to calculate the z score. Then, the standard normal distribution tables in Appendix F are used to determine the probabilities of finding scores higher or lower than calculated z scores. The formula to calculate $Z$ is:

$$z = \frac{x - \mu}{\sigma}$$

where $z$ = the standardised score, $x$ = the single score, $\mu$ = mean of the population and $\sigma$ = standard deviation.

For the single score of 48%, mean of 58% and standard deviation of 10%, the $Z$ score computes as:

$$z = \frac{48 - 58}{10} = -1.00$$

Ignore the minus sign. Looking at the values in the table in Appendix F, the probability $z$ score of 1.0 is 0.1587, thus indicating that 15.87% of scores in a normal distribution would be expected to be lower than 48%. If the calculation were repeated for the score of 68%, this would similarly give a $z$ score of 1.0, indicating 15.87% of scores in a normal distribution would be expected to be higher than 68%. The percentage of scores expected to be between 48% and 68% would thus be $100 - (15.87 + 15.87) = 68.26\%$.

Take the arbitrary figure of 40%; the computation is:

$$z = \frac{40 - 58}{10} = -1.80$$

Looking at the values in the table in Appendix F, the probability $z$ score of 1.80 is 0.0359, thus indicating that 3.59% of scores in a normal distribution would be expected to be lower than 40%. If the pass mark in an exam was 40%, this would indicate that 3.59% of students may be expected to fail.

## Confidence intervals

Confidence intervals are potentially an added aspect of complexity. If you can manipulate the formula, and show understanding of the principles involved (some of which need to be derived from inferential statistics in chapter 8), this gives you the opportunity to gain extra marks. Its application is in the calculation of mean scores; and remember mean scores are part of the descriptive analysis, if you are performing difference in mean tests or correlations.

It is often the case that mean scores are determined for samples; the sample size may be just 30, with a population size much larger than that. The whole sampling strategy hangs on the premise that results from the sample reflect the results that would have been achieved, had it been possible to survey the whole population. In Table 7.4, the mean score is 54.27%; are we confident that if we were able to survey the whole of the population, we would get that same mean score? The answer is no.

If the significance level set for the main study objective is $\leq 0.05$, can we be 95% confident we would get the same mean score? Answer, no. Therefore, what confidence intervals will do is set a range of scores where we can state that we are 95% confident (or another specified significance level), that the mean score of the population will lie within a calculated range. There is clearly a 5% probability the mean score will lie outside this range.

A narrow confidence interval range is welcome; this lends to the validity of the study. A wide confidence interval range severely limits validity; limited validity at undergraduate level is defensible, providing you recognise those limits. Confidence intervals help you to do this.

The key message with confidence intervals is that it is important in your discussion and conclusions, and in interpreting the validity of your own study, that you recognise your mean scores may not accurately reflect the population mean scores. Confidence intervals are calculated from the formula:

$$\bar{\mu} = \frac{\pm t \cdot (s) + \bar{X}}{\sqrt{n}}$$

Where $t$ = critical value at $\alpha = 0.05$ with a $\pm$ value, $\bar{X}$ = sample mean, $\bar{\mu}$ = population mean, s = standard deviation, $n$ = sample size.

It can be seen from the formula that small standard deviation and high $n$ will reduce the range. The critical $t$ value is obtained from the table in appendix K, critical values of the $t$ distribution. It is the size of $t$ resulting from a $t$-test calculation that cuts off the tails of the normal distribution; $t$ values equal to or greater than the stated values are significant. It is noted that $t$ becomes smaller as $n$ increases; this gives further weight to the principle that higher sample sizes will give narrower confidence interval ranges.

Using the data from Table 7.4:

Step 1; $\sqrt{N} = \sqrt{30} = 5.48$

Step 2; $s/5.48 = 9.79/5.48 = 1.79$

Step 3; critical value of $t = 2.042 \times 1.79 = 3.65$

Step 4: statement of finding. 95% confident that the mean score of the population lies in the range of $54.27 \pm 3.65 = 50.62$ to $57.92$.

### General use of descriptive statistics

Table 7.4 illustrates how all eleven descriptive statistics can be presented succinctly at the bottom of a table; that is count ($n$), sum, mean, median, mode, standard deviation, variance, maximum, minimum, range, and confidence intervals. All these statistics are given for each of the eight questions, var1a to var1h, in the basket. Using software they are quick to calculate. It is not appropriate that all these data should generate discussion against each individual question; this would be tedious. However, you should browse these statistics, and if any results are different to the norm, the reasons for this may warrant investigation. If you are able to undertake a pilot study, browsing descriptive statistics for individual questions and comparing your data to a normal distribution, this may prompt you to make adjustments to the final data collection instrument. Most discussion should arise from the descriptive statistics for the overall measure of var1 percentage, in column 10.

All data do not lend themselves to presentation like examination scores or in frequency histograms. Also, in our surveys, we are often dealing with small $n$, and though it may be

**Table 7.8** Two examples of data that may be presented in tables or line diagrams: output in the UK construction industry in £millions (Office of National Statistics 2009) and tender price and cost price indices (BCIS 2010).

| | | Output in the UK construction industry | | BCIS indices; base 1985 = 100 | |
|---|---|---|---|---|---|
| | Year | £million, adjusted to 2005 prices | Year | Tender price indice | Cost price indice |
| 1 | 1999 | 91 746 | 1999 | 151 | 183 |
| 2 | 2000 | 92 683 | 2000 | 161 | 191 |
| 3 | 2001 | 94 269 | 2001 | 174 | 196 |
| 4 | 2002 | 98 520 | 2002 | 187 | 205 |
| 5 | 2003 | 104 013 | 2003 | 197 | 215 |
| 6 | 2004 | 107 852 | 2004 | 213 | 227 |
| 7 | 2005 | 107 007 | 2005 | 224 | 241 |
| 8 | 2006 | 108 364 | 2006 | 230 | 255 |
| 9 | 2007 | 110 952 | 2007 | 245 | 267 |
| 10 | 2008 | 109 716 | 2008 | 245 | 282 |
| Count | | 10 | | 10 | 10 |
| Sum | | 1 025 122 | | 2027 | 2262 |
| Mean | | 102 512 | | 202.7 | 226.2 |
| Median | | 105 510 | | 205 | 221 |
| Mode | | – | | 245 | n.a. |
| Standard deviation | | 7489 | | 34.0 | 34.0 |
| Variance | | – | | – | – |
| Maximum | | 110 952 | | 245 | 282 |
| Minimum | | 91 746 | | 151 | 183 |
| Range | | 19 206 | | 94 | 99 |
| Confidence interval | | – | | | |

desirable that the distribution of numbers replicate a normal distribution, it may be that they do not. Even in these cases, descriptive statistics are often still appropriate. Line diagrams, pie charts, box-plots or scatter diagrams may sometimes better visualise data than frequency histograms.

Table 7.8 illustrates two pieces of secondary data that may be best presented by table and line diagram: (a) output in the UK construction industry (Office for National Statistics 2009), and (b) tender price and cost price indices (BCIS 2010). The data set for output is considered alone; the data set for indices illustrates how paired data from tables can be brought together for comparison. The eleven descriptive statistics are considered for each. For both data sets it is thought that mode, variance and confidence intervals have no real meaning; therefore, they are not computed. For construction industry output, it is easier from Figure 7.4 than from Table 7.8 to identify the steady growth in the period 2001–2004, and then two years of negative growth. For tender and cost price indices, it is easier from Figure 7.5 than from Table 7.8 to identify the steady and almost parallel increases until 2008, when costs continued to rise but tender prices levelled off. Discussion can take place around both the descriptive figures in the tables, and the trends that are most easily identified by the line diagrams.

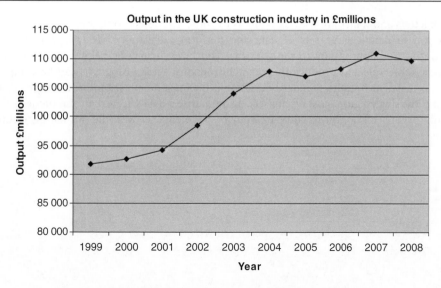

**Figure 7.4** Example of single data in a table illustrated in a line diagram.

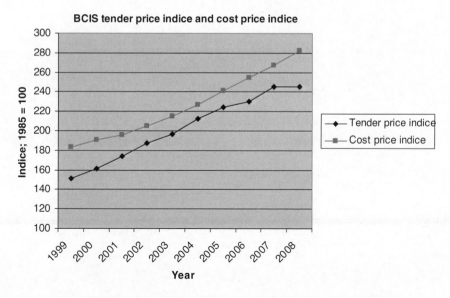

**Figure 7.5** Example of paired data in a table illustrated in a line diagram.

## Summary of this chapter

It is likely that, even if your study is predominantly qualitatively based, you will collect some statistical data, if only in the literature review. Descriptive statistics are frequently used in industry and academia to summarise large data sets with much fewer numbers and diagrams.

Greek letters and italics are used as symbols to abbreviate tests. Computer spreadsheets will help you to handle large sets of data and perform calculations. You should select the descriptive tool appropriate to the data that you have. It may be helpful to arrange data into class intervals and produce frequency histograms; these histograms can be compared with normal distributions. Standard deviation is the most often used measure of spread. The mean is the single figure most often used to summarise a data set; median and mode are also useful. $Z$ scores allow you to compare your bespoke data set with the normal distribution curve.

# 8 Quantitative data analysis: inferential statistics

The objectives of each section of this chapter are:

8.1 Introduction: to provide context for inferential statistical data collection and analysis
8.2 Probability values: to explain the *p* values concept
8.3 The chi-square test: to explain and illustrate manual calculations
8.4 Difference in means tests: to introduce the '*t*' test, and the concepts of unrelated/related data and parametric/non-parametric data. To help decide which type of *t* test be used
8.5 Difference in means: the unrelated Mann–Whitney test; to explain and illustrate manual calculations
8.6 Difference in means: the related Wilcoxon test; to explain and illustrate manual calculations
8.7 Difference in means: the parametric related *t* test; to explain and illustrate manual calculations
8.8 Correlations; to explain and illustrate manual calculations
8.9 Difference in means, correlations or both?: to answer, illustrated with an example
8.10 Using correlation coefficients to measure internal reliability and validity: to illustrate with example
8.11 Summarising results: to illustrate how to summarise

## 8.1 Introduction

This chapter builds on the descriptive statistical analysis in Chapter 7. If you are able to understand and apply concepts in this chapter to a quantitative study, you give yourself the possibility of gaining higher marks. Inferential statistical tools allow 'inferences' to be made about characteristics of populations based upon much smaller samples. They also contribute to theory building, establishing cause and effect relationships between variables.

*Writing a Built Environment Dissertation: Practical Guidance and Examples.* Peter Farrell.
©2011 Peter Farrell. Published 2011 by Blackwell Publishing Ltd.

## 8.2 Probability values

Chapter 4 explains that the way we measure things can be put on a continuum of (a) categorical (nominal), (b) ordinal, (c) interval, and (d) ratio. These are levels of measurement. The sequence is important: the richness of the data increases as the scale moves from (a) to (d). Similarly, some of the statistical tests can be put on the continuum of (a) chi-square, (b) difference in means, and (c) correlations. The richness and power of the tests increase as they move from (a) to (c). These tests are inferential tests; for a given set of data it is necessary to perform only one of the tests. For all tests, there needs to be two variables; an independent variable (IV) and a dependent variable (DV). One issue that drives which type of test can be used is the level of the data. For example if the IV and DV were at the categorical level, the appropriate test would be the chi-square; if the IV and DV were at the interval level, the appropriate test would be the correlation. Indicatively, if the level of the data is richer, the possibility of using richer tests is greater. In the same way that interval data give greater insight than categorical data, correlation tests may give greater insight than chi-square tests. However, in some studies, it is only appropriate to measure variables at the categorical level, and therefore also appropriate that the chi-square test is undertaken.

For each of the three types of test there are two stages in the analysis and two outputs at the end for each which comprise the results. The chi-square test has a contingency table and then a '$p$' value. The difference in means test produces mean scores and then a '$p$' value. A correlation test has a correlation coefficient and then a '$p$' value. So that you do not under-emphasise the first output of each test, it may be useful to think of it as the primary output (that is, depending on the type of test, the contingency table or the mean score or the correlation coefficient) and the '$p$' value as the secondary output.

Statistics textbooks seem to create a lot of fuss around the concept of probability or '$p$' values, also known as significance levels. Arguably, as a concept, it is made overly complex, and as a secondary output, inappropriately often takes precedence over the primary output. Students sometimes only report their '$p$' values in results sections, and do not report or comment on the primary contingency tables, mean scores or correlation coefficients; nor do they report on the direction of any relationships. It is as though studies are completely overtaken by '$p$' values, and Utopia is achieved if '$p$' is found to be $\leq 0.05$ or some other figure. You need to steer away from this possibility. Whichever test you may use, you must present your results with the primary contingency table, or mean scores, or correlation coefficient AND then in all cases, the '$p$' value.

The '$p$ value' or probability value is about 'chance'. To emphasise, chance, chance and chance. As a number, it always lies between zero and $+1.00$; it does not have negative values. When people take risks, there may be a five in 100 chance of a particular event occurring. Five in 100 is the same as '$p \leq 0.05$'. In the chi-square, difference in means and correlations tests, we are testing for links between the IV and the DV; that is, does the IV have an association with or causal effect on the DV. In the three tests, the first stage of looking for links is to 'eyeball' the data. In the chi-square test, eyeball the contingency table; you will be able to observe whether the data set is unbalanced, and therefore whether it appears that the IV is influencing the DV. In the difference in means test, eyeball the mean scores; if there is a large difference in mean scores, this may suggest that the IV influences the DV. In the correlation test, eyeball the scatter diagram; if the data appears as though it could be surrounded by a thin or narrow rugby ball, this may suggest that the IV influences the DV. However, the spread of the data may make it difficult to assert with confidence that the IV influences the DV. It may be a close call. Therefore, the second stage of the analysis in each case is to perform the statistical test and calculate the '$p$' value, to determine whether to the extent that there may be a relationship, did it occur due to chance?

### The p value of $\leq$ 0.05

The significance level is often set at $\leq 0.05$. At the end of your calculation you will get a $p$ value. If your calculation is undertaken by computer you will get a precise $p$ value figure to several decimal places. Alternatively, if your calculation is performed manually, you will compare the penultimate figure in the calculation to figures in tables to determine whether your $p$ value is $\leq$ 0.05. If the $p$ value is $\leq 0.05$, you will be able to assert that you 'reject the null hypothesis, the IV does influence the DV, with $p$ set at 0.05'. It may help that you think of this as you 'reject the null or the notion that there is no association or no difference or no relationship between the IV and the DV; there is cause and effect between the IV and the DV, and you are 95% sure of that'. Alternatively, the $p$ value may be $> 0.05$. Therefore, you will assert that you 'cannot reject the null hypothesis, the IV does not influence the DV, with $p$ set at 0.05'. It may help that you think of this as you 'cannot reject the null or the notion that there is no association or no difference or no relationship between the IV and the DV; it may be that there is no cause and effect between the IV and the DV'. If the $p$ value $> 0.05$, some students write words such as 'accept the null hypothesis'. By convention these types of words should not be used. Statistical tests are undertaken against the null and the finding is either, 'reject the null' or 'cannot reject the null'

If you find the $p$ value is $\leq 0.05$, and assert that you 'reject the null hypothesis, with $p$ set at 0.05', you do need to recognise that there is a five in 100 *chance* that you could be wrong. The numbers in your calculations could be such that, to the extent that a relationship between the IV and DV is found, it could be a fluke occurrence. If a subsequent study found that your claim for a link between the variables is incorrect, you will have made a 'type 1 error'. You need to recognise in your conclusions that you could be making a 'type 1 error'. Alternatively, if the $p$ value $> 0.05$ and you assert you 'cannot reject the null hypothesis', there is a possibility that a subsequent study may find there is a link between the variables. In this case you have made a 'type 2 error'.

The statistical calculations can only manipulate the raw numbers; whether the $p$ value that is 'spat out' at the end of the calculations is 0.05 or some other number, is derived solely by the original raw numbers. The factors that drive the $p$ value result are (a) sample size, (b) differences between numbers, and (c) the spread or distribution of the numbers. You need to keep in mind that there are other issues about validity in your study that will give emphasis to the weight you can place on any results in discussion and conclusion chapters.

If you find $p \leq 0.05$ in the context of one study, this is not a definitive link; you should recommend more work to test and re-test, with perhaps slightly different methodologies. If the result of your study is that the $p$ value is $> 0.05$ and you 'cannot reject the null hypothesis', it is reasonable that you do not suggest that there is definitely no link between the variables. It is correct that you have not found a link in your data. However, you must have been pointed in some way towards the possibility of a link by your own experience or the literature. You may therefore make a recommendation to re-test, using a larger sample size or slightly different methodology. Only if the testing, re-testing and re-re-testing by others does not identify a link, should 'we' abandon that exploration and move onto other potential links to the DV.

### Setting the significance level of p; alternatives to 0.05

The terms '$p$ value' and 'significance level' are sometimes used interchangeably and together. In the media, it may be reported that a particular study has achieved significant results. What lies behind the words 'significant results' is that statistical testing has been undertaken and $p$ has found to be $\leq 0.05$ or perhaps another figure. 'Significant results' is a phrase that the general public may loosely associate itself with; it simplifies the academic concepts and language, and avoids needing to report to the public that $p$ has been found to be $\leq 0.05$.

You should distinguish between significance levels and $p$ values. At the start of your study you should set your significance level; that is, set it at the outset of your study, and state it in the introduction. You will benchmark your study against the significance level that you set. This is sometimes also called 'setting alpha'. Lots of students blindly get 'hung-up' around the 0.05 level. It does not have to be $\leq 0.05$. This 0.05 is a level followed by convention in much social science work. It is supported by Siegel and Castellan (1988, p.9), but they also state that levels of significance should be set to reflect the 'perceived consequences of the application of the results'. Alternative cut-off points used by convention are:

$p \leq 0.001$: equivalent to a 1 in 1000 chance
$p \leq 0.01$: equivalent to a 1 in 100 chance
$p \leq 0.10$: equivalent to a 10 in 100 or 1 in 10 chance

The first two are widely used in some professions. In medical work, for example, with life or death decisions a recurring issue, '$p \leq 0.001$' may be appropriate. In construction, the issues are often not of such importance and '$p \leq 0.05$' may be acceptable. It may be only about a business decision to invest money, and companies may be willing to risk relatively small sums with only a five in 100 chance of getting it wrong. They may, however, be prepared to invest a larger sum with only a one in 100 chance of getting it wrong, and therefore in business $p$ may be set at $\leq 0.01$. In undergraduate dissertations, it is unlikely that the 'perceived consequences of the application of the results' is high. It unlikely that construction companies will invest substantial sums based solely upon your work, or take decisions that may influence issues such as health and safety. Therefore, it may seem reasonable at the start of your study that you set $p$ at $\leq 0.10$. If you do this, you need to substantiate that the 'perceived consequences of the application of the results' do not pose business or other risks. Most textbooks are silent on the possibility of setting $p$ at $\leq 0.10$; it is often 0.05 or nothing.

There are issues about why 0.05 is the often used cut-off point. Why not 0.03 or 0.08 or 0.15? Matthews (1998) states that 0.05 was selected by Fischer in the 1920s because it was 'convenient'; and that is that. There was a proposal by the British Psychological Society in 1995 to re-examine the use of $p$ values, but it was discarded because it was thought it 'would cause too much upheaval' for the academic community. $P$ values are used in all countries, in all disciplines; they are universal.

At some point there has to be a decision, such as (a) to invest or not to invest the money, (b) to change or not to change the system of work, or (c) to jump or not to jump into the river.

As Kerlinger stated, 'everything is either 1 or 0', therefore we often choose to jump at $p \leq 0.05$. However, even after you have set your significance level, be it at $\leq 0.01$, or $\leq 0.05$, or $\leq 0.10$, this still leaves the possibility of some 'greyness'. Imagine a study that was set up to test against a significance level of 0.05, and the outcome would influence a business decision to invest money. Suppose the $p$ value results were just over 0.05, say 0.06 or even up to 0.09. A narrow view would be to reject the null hypothesis, the IV does not influence the DV; do not invest the money. However, a business may want to look more closely at the result. There may be lots of potential advantages if the investment is successful. When you write up your findings, discussion and conclusions, you cannot retrospectively change the significance level from 0.05 to 0.10; this would be cheating. However, a $p$ value that is close to the established significance level will have substantial impact on the way that you write-up the latter part of your study, including your recommendations. If an experimental study were set up with $\leq 0.10$, and the result was 0.11, that $p$ value of 0.11 gives you greater incentive to continue study in this area than if a much higher $p$ value, say 0.40, were found.

At the end of your calculations you should state the $p$ value, and then whether it is equal to or lower than the set significance level. For example, if the result of a statistical test is $p = 0.04$, you should assert that: $p$ was calculated at 0.04 and with the significance level set at $p \leq 0.05$, the null hypothesis is rejected'. Whilst you may report the actual figure, the headline issue is whether or not $p$ is $\leq 0.05$.

If you set the significance at $p \leq 0.05$, and your result also is within the $p \leq 0.01$ cut off point, you may state $p$ was calculated at 0.009 and with the significance level set at $p \leq 0.05$, the null hypothesis is rejected. The result is also significant at the conventionally set cut-off point of $p \leq 0.01$. The headline issue remains that you reject the null hypothesis with $p$ set at $\leq 0.05$. A significant difference is not an indicator of the size effect, or on the importance of the difference. Judgments on the importance of statistical differences must be made subjectively. A difference between two means of 10% with a small sample size may not be so important as a difference in means of 5% with a larger sample size, even though both data sets may have $p$ calculated at $\leq 0.05$. If you were to perform an inferential test, using two identical data sets (though in practice you would not do this), the calculated $p$ value will be 1.00. The more two data sets become dissimilar, the more $p$ moves away from 1.00 towards 0.05 and below.

## 8.3 The chi-square test

The Pearson chi-square test is often just called the chi-square test, without Pearson's name. It is 'square' not 'squared'. Chi is pronounced as 'Ki', as though rhyming with fly. Sometimes it is represented by the Greek letter $\chi$. Added to $\chi$, is 'squared', expressed thus $\chi^2$, spelt out, chi-square. It compares the actual distribution of a set of frequency counts with theoretical or expected frequency counts. If the actual frequencies are 'significantly' different to the expected frequencies, it is inferred that the difference is due to an association between the variables; the null hypothesis will be rejected, and it is found there is an association between the IV and the DV. If the differences between actual frequencies and expected frequencies are small, it is inferred there is no association between the variables; the null hypothesis cannot rejected. In this latter case, to the extent that there are differences between actual frequencies and expected frequencies, these differences occur due to chance.

This test is used if both the IV and DV are measured at the categorical/nominal level. Another often-used label by statisticians, which can be helpful in thinking through the principles involved, is 'groups'; both the IV and DV have 'values', which can be placed into groups. A contingency table is drawn to illustrate the layout of the data. The table may be a simple $2 \times 2$, or if measures are available in more than two categories, it may be larger, e.g. $2 \times 3$, $3 \times 3$, etc. As contingency tables get too large, it makes interpretation of the results more difficult. A $2 \times 2$ is often adequate for undergraduate work, but perhaps a $2 \times 3$ is also appropriate. Single row chi-square tests are also possible.

The raw data are best placed firstly in a spreadsheet. Since spreadsheets lend themselves more readily to numbers, not words, it is appropriate to code each 'group' (or category) with a number, starting with the number 1 (not zero). Table 8.1 illustrates a raw data set for a

**Table 8.1** Raw data for a $2 \times 2$ chi-square.

| 1 Project number | 2 The IV; method of procurement. Competitive bidding = 1, collaborative working = 2 | 3 The DV; cost predictability. On or below budget = 1, over budget = 2 | 4 Contingency table cell A = 1,1 B = 1,2 C = 2,1 D = 2,2 | 5 Ditto, also enumerated |
|---|---|---|---|---|
| 1 | 2 | 2 | D | D1 |
| 2 | 2 | 1 | C | C1 |
| 3 | 1 | 1 | A | A1 |
| 4 | 1 | 1 | A | A2 |
| 5 | 1 | 2 | B | B1 |
| 6 | 2 | 2 | D | D2 |
| 7 | 1 | 2 | B | B2 |
| 8 | 1 | 2 | B | B3 |
| 9 | 1 | 2 | B | B4 |
| 10 | 2 | 2 | D | D3 |
| 11 | 2 | 1 | C | C2 |
| 12 | 1 | 2 | B | B5 |
| 13 | 2 | 1 | C | C3 |
| 14 | 1 | 1 | A | A3 |
| 15 | 2 | 1 | C | C4 |
| 16 | 1 | 1 | A | A4 |
| 17 | 2 | 1 | C | C5 |
| 18 | 1 | 1 | A | A5 |
| 19 | 1 | 2 | B | B6 |
| 20 | 1 | 1 | A | A6 |
| 21 | 2 | 1 | C | C6 |
| 22 | 2 | 1 | C | C7 |
| 23 | 1 | 2 | B | B7 |
| 24 | 1 | 2 | B | B8 |
| 25 | 1 | 2 | B | B9 |
| 26 | 1 | 2 | B | B10 |
| 27 | 1 | 2 | B | B11 |
| 28 | 2 | 1 | C | C8 |
| 29 | 2 | 1 | C | C9 |
| 30 | 2 | 1 | C | C10 |
|  | 1 = 17No, 2 = 13No | 1 = 10No, 2 = 3No | | A = 6 |
|  | | | | B = 11 |
|  | | | | C = 10 |
|  | | | | D = 3 |

hypothetical study with an IV and DV. It is based on the assumption that you are a part-time student who works for a client that procures buildings from the construction industry. Sometimes you invite bids for projects competitively, and sometimes by frameworks/ partnering/collaboration. From the perspective of clients, the objective is to 'determine whether method of procuring work influences cost predictability'. As a null hypothesis for testing, this is written 'method of procuring work does not influence cost predictability'. The significance level is set at $p \leq 0.05$. The IV is a method of procuring work. The method of procuring work is in two 'groups' or it has two 'values': (a) competitive bidding, coded number 1, and (b) collaborative working, coded number 2. The DV is cost predictability measured on completion of projects. It is placed in two groups or has two values: (a) completion on or below budget, coded number 1, and (b) completion over budget, coded number 2. On the one hand the study is intended to be inferential, so that results could be inferred to replicate the whole population. However, a sample of convenience is taken, comprising data from completed projects in your own organization. This sample of convenience effectively classifies the research as a case study; this is fine, providing its limitations are recognised.

Data are obtained from 30 completed projects, selected randomly. Thirty is the sample number, often referred to as '$n$'. If it is possible to get data for more, you should do so. It is not necessary to get equal numbers of 15/15 for the IV in each group, but the more closely the numbers are to each other the better. Sixteen in one group and 14 in the other is fine; 17/13, 18/12 is also fine. If the balance starts to tip to 19/11 or 20/10, the sample size of 10 for one group is arguably becoming too low. Group sizes of 25/5 would be completely inappropriate.

### Assembling the raw data

Use a spreadsheet to assemble and sort the raw data; number each of the projects, 1 to 30 in column 1 (Table 8.1). For columns 2 and 3, label each project based on its method of procurement and whether it was completed on budget. Then, in preparation for transferring the data into a 2 × 2 contingency table, in column 4, allocate each project to one of four cells. A project can be either:

- Competitive bid, and completed on or below budget, coded 1,1
- Competitive bid, and completed over budget, coded 1,2
- Procured by collaboration, and completed on or below budget, coded 2,1
- Procured by collaboration, and completed over budget, coded 2.2
- Codes 1,1 are allocated to cell A (top left cell)
- Codes 1,2 are allocated to cell B (top right cell)
- Codes 2,1 are allocated to cell C (bottom left cell)
- Codes 2,2 are allocated to cell C (bottom right cell)

### Transferring raw data to the contingency table: stage 1

The contingency table is illustrated in Table 8.2, cell letters only. Table 8.3 illustrates the numbers, transferred from Table 8.1, into the cells of the contingency table.

**Table 8.2** Cell labels for a chi-square contingency table.

| | | Cost predictability | | |
| --- | --- | --- | --- | --- |
| | | On or below budget | Over budget | Totals |
| Method of procurement | Competitive bidding | Cell A | Cell B | Total row 1 |
| | Collaborative working | Cell C | Cell D | Total row 2 |
| | Totals | Total column 1 | Total column 2 | Overall total |

**Table 8.3** Numerical values in a chi-square contingency table (based upon data in Table 8.1).

| | | Cost predictability | | |
| --- | --- | --- | --- | --- |
| | | On or below budget | Over budget | Totals |
| Method of procurement | Competitive bidding | 6 | 11 | 17 |
| | Collaborative working | 10 | 3 | 13 |
| | Totals | 16 | 14 | 30 |

As part of the stage 1 analysis, eyeballing the data in Table 8.3 suggests that the IV may be influencing the DV; method of procurement influences cost predictability. It is important that you write about the direction of the relationship; is it competitively bid or collaborative working projects that are completed on or below budget? The data show that the majority of collaborative working projects (10 of 13) are completed on or below budget. The majority of competitively bid projects (11 of 17) are completed over budget.

If you feel unable to go to stage 2 of the analysis, just do stage 1. If you can write in your dissertation to illustrate your understanding of the contingency table in stage 1, this may be sufficient to achieve a pass grade; but talk to your supervisor about it. However, more marks are likely to be awarded if you are able to complete stage 2 of the analysis.

### Are differences due to chance? manual calculations: p values and degrees of freedom

Stage 2 determines whether, to the extent that there are differences between the numbers in the contingency table, those differences could be due to *chance*? It is expressed in a mathematical formula, thus:

$$\text{Chi-square} = \frac{\Sigma(\text{observed frequencies} - \text{expected frequencies} - 0.5)^2}{\text{expected frequencies}}$$

where '$\Sigma$' is the Greek letter sigma, representing 'the sum of'.

For simplicity, 'stage 2' can be separated into 'six steps' in a manual calculation, summarised in Tables 8.4 and 8.5. A seventh step is a statement of the finding, including

**Table 8.4** Expected frequencies (if the IV did not influence the DV).

| | | Cost predictability | | |
| --- | --- | --- | --- | --- |
| | | On or below budget | Over budget | Totals |
| Method of procurement | Competitive bidding | A 9.06 | B 7.93 | 17 |
| | Collaborative working | C 6.93 | D 6.06 | 13 |
| | Totals | 16 | 14 | 30 |

**Table 8.5** Five of the six steps in stage 2.

| Cell | Step 1: calculate the expected frequencies in each cell, by 'row total, multiplied by column total, divided by overall total' | Step 2: deduct expected and actual frequencies from each other; no negative signs | Step 3: take step 2 answers, and deduct 0.50 | Step 4: take step 3 answers, and square them; then divide by expected frequency |
| --- | --- | --- | --- | --- |
| A | $17 \times 16/30 = 9.06$ | $9.06 - 6 = 3.06$ | $3.06 - 0.50 = 2.56$ | $2.56^2/9.06 = 0.72$ |
| B | $17 \times 14/30 = 7.93$ | $7.93 - 11 = 3.07$ | $3.07 - 0.50 = 2.57$ | $2.57^2/7.93 = 0.83$ |
| C | $13 \times 16/30 = 6.93$ | $6.93 - 10 = 2.07$ | $2.07 - 0.50 = 1.57$ | $1.57^2/6.93 = 0.35$ |
| D | $13 \times 14/30 = 6.06$ | $6.06 - 3 = 3.06$ | $3.06 - 0.50 = 2.56$ | $2.56^2/6.06 = 1.08$ |

Step 5: take step 4 answers, and add them together to obtain the chi-square value  2.98

'direction'. Table 8.4, illustrates the expected frequencies. Table 8.5 presents five of the first six steps in tabular form, including calculations for the expected frequencies in Table 8.4. The first of the six steps, and the principle underpinning the chi-square, is to calculate the frequency counts in each cell that would have been expected if the IV does not influence the DV; and then compare the actual frequencies (your data) with the expected frequencies. If there are small differences between the actual and expected frequencies, the sixth and last step of the calculation will show the $p$ value higher than your set significance level; your finding will be 'cannot reject the null hypothesis'. Alternatively, if there are large differences between the actual and expected frequencies, the sixth step will show the $p$ value lower than your set significance level; your finding will be 'reject the null hypothesis'.

The 'formula' for calculating the expected frequencies in each cell, as step 1, is simply: 'row total, multiplied by column total, divided by overall total'. Thus for cell 'A' the row 1 total is 17, column 1 total 16, and for all cells the overall total is 30. Therefore, the expected frequency for cell 'A' is $17 \times 16/30 = 9.06$. As a theoretical number, it is usual to express it to two decimal places. Expected values for all cells, including 'B', 'C' and 'D' are shown in Table 8.4.

The six steps are:

(1)  Step 1: calculate the expected frequencies in each cell, by 'row total, multiplied by column total, divided by overall total'.
(2)  Step 2: deduct expected and actual frequencies from each other; no negative signs.

(3) Step 3: take step 2 answers, and deduct 0.50. Deducting 0.50 is known as Yates' correction for continuity. Yates' correction is also used if the expected frequency in one cell or more (not actual frequency) is less than five. Not all authors agree that Yates' correction should be applied; they think it over-corrects. Its effect is to reduce the chi-square value and thus increase the $p$ value. Note, the chi-square value in Table 8.5, without Yates' correction, would be $3.06^2/9.06 + 3.07^2/7.93 + 2.02^2/6.39 + 3.06^2/6.06 = 1.03 + 1.18 + 0.61 + 1.54 = 4.36$.

(4) Step 4: take step 3 answers, and square them; then divide by expected frequency.

(5) Step 5: take step 4 answers, and add them together to obtain the chi-square value.

(6) Step 6: evaluate the calculated chi-square value with the critical values of chi-square in the table in Appendix G; state whether the finding is to reject or not to reject the null hypothesis.

In step 6, the operation of the chi-square table is based on the chi-square value in your calculation being equal to or exceeding the pre-determined critical values in the table in Appendix G. These values were determined as part of the original design of the test by Karl Pearson. To use the table requires an explanation about the concept of degrees of freedom. The range of chi-square values in the table are given in rows for varying numbers of degrees of freedom. A degree of freedom can be defined as:

'total number from a sample which has to be known, when the overall total is known, to be able to determine the missing data'

To illustrate this, Table 8.3 is represented as Table 8.6, but with data from cells 'B', 'C' and 'D' missing. Given the data in the table, we must '. . . determine the missing data', that is, the numbers that go in cells 'B', 'C' and 'D'. Cell 'B' must be 11 (17–7). If cell 'B' is 11, cell 'D' must be 3 (14–11). If cell 'A' is 6, cell C must be 10. Therefore, with only one 'number from the sample . . . known', that is cell 'A' at 6, and the overall totals known, that is 17, 13, 16 and 14, it is possible to determine the missing data, that is, the data in cells 'B', 'C' and 'D'. The degrees of freedom for a 2 × 2 contingency table is always 1.

Table 8.7 illustrates a 2 × 3 table, where the DV, cost predictability, has three values or is in three groups: (a) on or below budget, (b) up to 10% over budget, and (c) more than 10% over budget. In a 2 × 3 table, following the explanation for table 8.6 through, the degrees of freedom is always 2. The total number from the sample which has to be known is 2; in this case, cells 'A' and 'B' are known to be able to determine the missing data in cells 'C', 'D', 'E'

**Table 8.6** Degrees of freedom in a 2 × 2 table.

| | | Cost predictability | | |
|---|---|---|---|---|
| | | On or below budget | Over budget | Totals |
| Method of procurement | Competitive bidding | 6 | Cell B | 17 |
| | Collaborative working | Cell C | Cell D | 13 |
| | Totals | 16 | 14 | 30 |

**Table 8.7** Degrees of freedom in a 2 × 3 table.

| | | Cost predictability | | | |
|---|---|---|---|---|---|
| | | On or below budget | Up to 10% over budget | More than 10% over budget | Totals |
| Method of procurement | Competitive bidding | 6 | 5 | Cell C (6) | 17 |
| | Collaborative working | Cell D (10) | Cell E (3) | Cell F (0) | 13 |
| | Totals | 16 | 8 | 6 | 30 |

and 'F'. Rather than draw up a table and think the logic through if you have contingency tables larger than 2 × 3, the formula for chi-square is:

degrees of freedom = (number of rows − 1) × (number of columns − 1)
For a 2 × 2 table: $(2 − 1) × (2 − 1) = 1$
For a 2 × 3 table: $(2 − 1) × (3 − 1) = 2$

For the stated significance level of $p \leq 0.05$, the chi-square value in the table in Appendix G, for 1 degree of freedom, is 3.84. Our value of 2.98 does not exceed 3.84; we cannot reject the null hypothesis with the significance level set at $\leq 0.05$. Note that if the significance level had been set at $\leq 0.10$, it would have been significant.

In Step 7 formally state the finding and direction. The null hypothesis cannot be rejected with $p$ set at $\leq 0.05$; method of procurement does not influence cost predictability. Differences that are noted by eyeball observation are not significant.

### The consequence of larger sample size and different spread of numbers

Table 8.3 illustrates data that may be real to life. To illustrate the consequence of a larger sample size and a different spread of numbers, consider some numbers that are unlikely to occur in real life. Table 8.8 has equal numbers in each group for the IV and DV; 100 in each. It is difficult to tell by eyeballing the data whether the IV influences the DV. The $p$ value, calculated by Excel, is 0.15. To the extent that there are differences, they could have occurred due to chance. Therefore, it is found that the null hypothesis cannot be rejected; the IV does not influence the DV.

Now consider Table 8.9, which uses the same proportions from Table 8.8, but the frequency in each cell is increased tenfold. The $p$ value, calculated by Excel, is 0.0000077, or $p \leq 0.001$.

**Table 8.8** Smaller sample size, not significant differences, $p = 0.15$.

| | DV group or value 1 | DV group or value 2 | Totals |
|---|---|---|---|
| IV group or value 1 | 45 | 55 | 100 |
| IV group or value 2 | 55 | 45 | 100 |
| Totals | 100 | 100 | 200 |

**Table 8.9** Larger sample size, significant differences, $p = 0.0000077$ or $p \leq 0.001$.

|  | DV group or value 1 | DV group or value 2 | Totals |
|---|---|---|---|
| IV group or value 1 | 450 | 550 | 1000 |
| IV group or value 2 | 550 | 450 | 1000 |
| Totals | 1000 | 1000 | 2000 |

The effect of sample size is clear; since the numbers are larger, we can now be sure that the differences that are there have not occurred due to chance. It is found that the null hypothesis is rejected; the IV influences the DV.

Now consider Table 8.8 as the template with total frequency counts in the rows and columns as 100, and overall frequency count of 200, but adapted for Table 8.10. The frequencies in the cells are adjusted slightly, to show differences that have increased from 10 (45–55) to 20 (40–60). The $p$ value is reduced from $p = 0.15$ to $p = 0.004$ or $p \leq 0.01$. Consequently, in the latter case, with larger differences, it is found that the null hypothesis is rejected; the IV influences the DV.

Finally consider Table 8.9 as the template, with total frequency counts in the rows and columns as 1000, and an overall frequency count of 2000, but now adapted for Table 8.11. It does not have to be the case, to reject the null hypothesis, that the balance of frequency counts needs to be diagonally 'symmetrical'. In Table 8.11, the highest frequencies counts are both in column 2. There is high sample size, and there are noticeably 'different differences' in the rows $(550 - 450 = 100$ and $510 - 490 = 20)$; the $p$ value is 0.0012 or $p \leq 0.01$. It is found that the null hypothesis is rejected; the IV influences the DV.

## More complex or more simple chi-square

The $2 \times 2$ contingency table is called the simple chi-square. A contingency table larger than $2 \times 2$ is a complex chi-square. Table 8.7 uses a $2 \times 3$ contingency table to illustrate the principle of degrees of freedom. As a reminder, the DV, cost predictability, has three values or is in three

**Table 8.10** Smaller sample size, significant differences, $p = 0.004$ or $p \leq 0.01$.

|  | DV group or value 1 | DV group or value 2 | Totals |
|---|---|---|---|
| IV group or value 1 | 40 | 60 | 100 |
| IV group or value 2 | 60 | 40 | 100 |
| Totals | 100 | 100 | 200 |

**Table 8.11** Larger sample size, significant differences, frequency counts not diagonally symmetrical, $p = 0.00102$ or $p \leq 0.01$.

|  | DV group or value 1 | DV group or value 2 | Totals |
|---|---|---|---|
| IV group or value 1 | 450 | 550 | 1000 |
| IV group or value 2 | 490 | 510 | 1000 |
| Totals | 940 | 1060 | 2000 |

groups: (a) on or below budget, (b) up to 10% over budget, and (c) more than 10% over budget. It could be a $2 \times 4$ table if data were available to be able to classify the DV as (a) on or below budget, (b) up to 5% over budget, (c) more than 5%, up to 10% over budget, and (c) more than 10% over budget. The IV, method of procurement, could possibly be split into three or more groups for a $3 \times 3$ or $3 \times 4$ contingency table, since some projects may be awarded that have an element of competition and collaborative working in them; a half-way position, sometimes called competitive dialogue. The formula and calculation procedure for a complex chi-square is the same as the simple chi-square; remember the degrees of freedom will be higher.

A 'more' simple chi-square is the one row chi-square. The concept that is different in the one row chi-square is that it needs theoretical expected frequencies, ideally from the literature. Two examples are provided. Firstly, suppose data are available for completion of projects to budget; as a DV this is again expressed as 'cost predictability'. It can be in two groups, or have two values, e.g. (a) on or below budget, and (b) over budget. Alternatively, it can be in three or more groups as in Table 8.7. The objective is to compare projects completed in a sample, to industry norms; the sample may be projects in a single company (a case study) or in a sector of industry. To the extent that there may be differences between the sample and the industry norm, are those differences so large that they are 'significantly' different, i.e. have not occurred due to chance?

In the sample, data on 200 projects are available. Eighty are completed on or under budget, and 120 are completed over budget. Constructing Excellence in the Built Environment reports that on demonstration projects (those signed-up to the principles of collaborative working) 48% are completed on or under budget. The data are presented in Table 8.12, as though in a spreadsheet. The respective cells are:

- A1 = 80 projects actually completed on or under budget
- B1 = 120 projects actually completed over budget
- C1 = 96 projects ($200 \times 0.48$) that theoretically should have been completed on or under budget
- D1 = 104 projects ($200 - 96$) that theoretically would have been completed over budget

The formula and calculation procedure for the one row chi-square is the same as the simple chi-square; the degrees of freedom is the number of categories minus 1; for Table 8.12 this is $2 - 1 = 1$. The $p$ value, calculated by Excel, is 0.023. The findings are: reject the null hypothesis with $p$ set at $\leq 0.05$; the data from the sample are significantly different to industry norms published by Constructing Excellence; the performance in the sample is worse.

The second example of the one-row chi-square is based upon a traffic engineer who is interested in road design. Secondary data are available on the Internet about the number of motor bicycle accidents in a locality. The data are shown in Table 8.13. The actual number of accidents that have taken place are categorised by age of driver in seven age band categories; the degrees of freedom therefore is 6. Also on the Internet is an estimate of the age of the UK

**Table 8.12** One row chi-square, $p = 0.023$ or $p \leq 0.05$.

| | A | B | C | D |
|---|---|---|---|---|
| 1 | 80 | 120 | 96 | 104 |

**Table 8.13** Actual and expected frequencies of riders involved in motor bicycle accidents.

| | <16 | 16–19 | 20–29 | 30–39 | 40–49 | 50–59 | >60 | Total |
|---|---|---|---|---|---|---|---|---|
| (1) Age in years | | | | | | | | |
| (2) Actual number of accidents | 8 | 116 | 75 | 86 | 67 | 30 | 13 | 395 |
| (3) Percentage of motor cycling population | 1 | 9 | 10 | 31 | 28 | 13 | 8 | 100 |
| (4) Expected number of accidents in each age category given a total of 395 | 3.95 | 35.55 | 39.5 | 112.45 | 110.6 | 51.35 | 31.6 | 395 |

population of riders in the same age bands; for example 31% of riders are aged 30–39. Expected frequencies are calculated by percentage of riders in each category multiplied by the total number of accidents, e.g. age 30–39 $= 31\% \times 395 = 112.45$. The chi-square calculation compares the actual frequencies in row 2 with the expected frequencies in row 4. The $p$ value, calculated by Excel, is $\leq 0.05$. The findings are: reject the null hypothesis with $p$ set at $\leq 0.05$; the actual ages of riders involved in accidents are significantly different to expected ages; the data show that more young riders are involved in accidents.

The important element in the one row chi-square is that the theoretical frequencies should come from the literature, and ideally not an informal assumption or guess (e.g. 50%/50% split for a variable). If you do make an 'informal assumption', you should go as far as is possible to substantiate it.

## 8.4 Difference in mean tests: the '*t*' test

Tests that are sometimes called difference in mean tests are richer than the chi-square test. They actually do more than compare differences means; they compare the distribution of numbers in a data set. Mean scores, which are a key output from the tests, are more informative than contingency tables. One type of difference in means test, often referred to, is the '*t*' test. To use difference in means tests, one of the variables (the IV or the DV, it does not matter which) needs to be at the categorical/nominal level. The label 'groups', similar to the concepts of things being arranged into categories, is useful in thinking through principles about when to use difference in means tests. For the other variable (the IV or the DV), the level of measurement needs to be at the ordinal or interval level.

The principle of groups is often used in experiments, sometimes but not always, involving people. In such cases two (or more) groups are considered, and some 'treatment' is administered to one group. The two groups are called: (a) a control group that does not have treatment, and (b) an experimental group which is subjected to a treatment. Though strict ethical procedures have to be followed, in medical research the treatment may be some kind of drug (the IV is treatment, and it has two values; treatment or no treatment), and the variable measured for each group could be recovery from illness (the DV). A treatment may not have to be administered, since some groups are naturally occurring in society, e.g. group 'A' people who smoke, and group 'B' people who do not smoke, or group 'A' white collar workers, and group 'B' blue collar workers or the most naturally occurring, group 'A' males, and group 'B' females. Lots of work in health-related sectors are based upon data being analysed in groups. In construction, methods of work, sector of work, or type of materials can be classified in groups.

There are two issues of small complexity to consider in difference of means tests that do not appear in the chi-square test: (a) are the raw data 'unrelated' or 'related' and (b) are the raw data 'non-parametric' or 'parametric'. Examples follow to illustrate these issues. You will need to classify your data around both issues, such that your data may be:

- Unrelated and non-parametric
- Related and non-parametric
- Unrelated and parametric
- Related and parametric

The above is important since, alongside knowing the number of 'groups', it determines which statistical test is appropriate for use.

### Unrelated or related data

Unrelated data are collected 'singularly'; it is a collection of data or numbers that cannot be paired-up with any other number in the collection; all numbers are unrelated to each other. Related data are measures of the same variable, and can be of two kinds: (i) matched subjects; scores for a subject can be paired-up with scores for another subject, e.g. often used in medical studies involving twins, and (ii) repeated measures; two scores are obtained from every subject. Repeated measures may involve collecting data before and after in some kind of experiment; perhaps measures of change. Related data are highly valued since to obtain significant results, smaller sample sizes are required. This can be illustrated using the data set in Tables 8.22 and 8.23 later. The scenario is based on related data, and the $p$ value calculated using Excel is 0.002. If the Mann–Whitney test is applied to the same set of numbers (note here, it is the 'numbers', since the scenario is not appropriate for the Mann–Whitney test), the $p$ value is not so highly significant at 0.02.

If you misinterpret unrelated data for related data, or vice versa, you will have a fundamental (and catastrophic) error in your work. It is therefore important that you are clear on this issue; ideally you should be clear before you collect your data. If in doubt, ask for help.

The terms that are used here are unrelated and related. The language used by statisticians can be really unhelpful; different words or phrases are often used to describe the same concepts. We just have to accept this, and be able to use these words and phrases interchangeably. Alternative terms for 'unrelated', used by some authors, are unmatched, or different, or between subjects, or repeated measures; the term 'related' can alternatively be matched, or same, or within subjects. For emphasis, this is shown in Table 8.14. It is hoped that authors who may use the term 'unmatched', will use the term 'matched' for the alternative study.

**Table 8.14** Unrelated and related studies—alternative terms or phrases.

| Four other terms used to describe the related study | Matched | Same | Between subjects | Repeated measures |
|---|---|---|---|---|
| Three other terms used to describe the unrelated study | Unmatched | Different | Within subjects | |

## Determining whether the data set is parametric

Many undergraduate students do not attempt to classify their data as parametric or non-parametric; since there is some modest complexity in it, as an issue it is ignored. In such cases they may select a non-parametric test or parametric test without justifying their choice and without demonstrating insight into what they are doing. If you can address it as an issue, and illustrate that you have appropriate understanding, it gives you the opportunity to gain extra marks. If you ignore it, or get it wrong, it is arguably not catastrophic.

The word parametric is defined in dictionaries as 'boundaries'. In a statistical sense, do the data fall within certain boundaries? If they do, the data are parametric; if they do not, they are non-parametric. The three boundaries adapted from Bryman and Cramer (2005), are:

(1)    The data must be at the level of interval or ratio scales.
(2)    The distribution of the population scores must be nearly normal.
(3)    The standard deviation of the variables should be similar.

Chapter 4 illustrates how judgement can be made whether criterion (1) is met.

There are three ways to make a judgement about criterion (2). The first is to plot your data in a frequency histogram, and make an eyeball judgement. The second is indicative only, since the raw scores are diluted by arranging into class intervals. It draws on the calculation of $z$ scores and chi-square, as noted earlier in Chapter 7. Using $z$ scores, given your data set of size $n$, you can calculate the theoretical spread of your data set if it replicated exactly a normal distribution. Using the one row chi-square, it is then possible to compare your actual data set with the theoretical normal distribution. If your data set does not resemble a normal distribution (it is different to a normal distribution), the result of the test will be a $p$ value less than your stated significance level, e.g. $\leq 0.05$. If your data set does resemble a normal distribution, the result of the test will be greater than the stated significance level. The data set in Table 7.5 is adapted to illustrate this, in Table 8.15. The class intervals remain 5 points wide, but their position is changed to a classification around the mean score of 54.27%, e.g. instead of a class interval of 50–54%, the intervals are arranged lower and higher around the mean with a mid-interval of 52.27–56.27%. The mean score of 54.27% lies in the middle of 52.27–56.27%.

Z scores are calculated for the extreme points on each class interval, e.g. for the point 32.67, and standard deviation of 9.79, the calculation is:

$$z = \frac{32.67 - 54.67}{9.79} = -2.24 (\text{column 4})$$

Looking at the values in the table in Appendix F, the probability $z$ score of 2.24 is 0.0125 (column 5); thus indicating that 1.25% of scores in a normal distribution would be expected to be lower than a score of 32.67%. The theoretical frequency count is $0.0125 \times 30$ ($n$) $= 0.375$ (column 7). The theoretical expected frequencies in the class interval 36.27–32.27 are the percentage scores for each deducted from each other; $0.036 - 0.0125 = 0.0211$ (column 6) $\times 30 = 0.633$ (column 7).

The theoretical frequency count at the mean and mid-point of 54.27% is counted twice, since in the distribution, on the horizontal scale, there are 15.54% of scores to the left of the

**Table 8.15** Computation of theoretical frequency counts for a normal distribution, $n = 30$. Based upon the data in Table 7.5. Mean var1 $= 54.27\%$.

| | 1 Var1 sort ascendency | 2 Class interval | 3 Actual frequency count | 4 Z score for extreme points on the class intervals | 5 Probability score for each Z score | 6 Percentage of theoretical scores in class interval | 7 Theoretical scores frequency count |
|---|---|---|---|---|---|---|---|
| - | | Below 32.27 | 0 | -2.24 | 0.00<br>0.0125 | 0.0125 | 0.375 |
| 1 | 34 | 32.27 to 36.27 | 1 | -2.240<br>-1.838 | 0.0125<br>0.0336 | 0.0211 | 0.633 |
| 2 | 38 | 36.27 to 40.27 | 2 | -1.838<br>-1.430 | 0.0336<br>0.0764 | 0.0428 | 1.284 |
| 3 | 40 | | | | | | |
| 4 | 44 | 40.27 to 44.27 | 1 | -1.430<br>-1.021 | 0.0764<br>0.1562 | 0.0798 | 2.394 |
| 5 | 47 | 44.27 to 48.27 | 5 | -1.021<br>-0.612 | 0.1562<br>0.2743 | 0.1181 | 3.543 |
| 6 | 47 | | | | | | |
| 7 | 47 | | | | | | |
| 8 | 47 | | | | | | |
| 9 | 47 | | | | | | |
| 10 | 50 | 48.27 to 52.27 | 2 | -0.612<br>-0.204 | 0.2743<br>0.4207 | 0.1464 | 4.392 |
| 11 | 50 | | | | | | |
| 12 | 53 | 52.27 to 56.27 | 9 | -0.204<br>0.50<br>0.204 | 0.4207<br>0.50<br>0.4207 | 0.0793 + 0.0793 = 0.1586 | 4.758 |
| 13 | 53 | | | | | | |
| 14 | 53 | | | | | | |
| 15 | 53 | | | | | | |
| 16 | 53 | | | | | | |
| 17 | 53 | | | | | | |
| 18 | 56 | | | | | | |
| 19 | 56 | | | | | | |
| 20 | 56 | | | | | | |
| 21 | 59 | 56.27 to 60.27 | 3 | 0.204 | 0.4207<br>0.2743 | 0.1464 | 4.392 |
| 22 | 59 | | | | | | |
| 23 | 59 | | | | | | |
| 24 | 63 | 60.27 to 64.27 | 3 | 0.612 | 0.2743<br>0.1562 | 0.1181 | 3.543 |
| 25 | 63 | | | | | | |

*(continued)*

**Table 8.15**  (Continued)

| 1 Var1 sort ascendency | 2 Class interval | 3 Actual frequency count | 4 Z score for extreme points on the class intervals | 5 Probability score for each Z score | 6 Percentage of theoretical scores in class interval | 7 Theoretical scores frequency count |
|---|---|---|---|---|---|---|
| 26<br>- <br>63 | 64.27 to 68.27 | 0 | 1.021 | 0.1562<br>0.0764 | 0.0798 | 2.394 |
| 27<br>28<br>29<br>69<br>69<br>72 | 68.27 to 72.27 | 3 | 1.430 | 0.0764<br>0.0336 | 0.0428 | 1.284 |
| 30<br>75 | 72.27 to 76.27 | 1 | 1.838 | 0.0336<br>0.0125 | 0.0211 | 0.633 |
|  | Above 76.27 | 0 | 2.24 | 0.0125<br>0.00 | 0.0125 | 0.375 |
| totals |  | 30 |  |  | 1.00 | 30 |

mean and 15.54% to the right. Table 8.15 illustrates the percentage of theoretical scores in the class interval around the mean (52.27–56.27) is computed by 7.93% of scores below the mean between 52.27 and 54.27, and 7.93% of scores above the mean between 54.27 and 56.27.

The figures in columns 3 and 7 are represented in a chi-square contingency table, in Table 8.16, showing actual and theoretical frequencies. The histograms for each are shown in Figures 8.2a and 8.2b. The $p$ value in the chi-square calculation is 0.348. It is found: cannot reject the hull hypothesis. To the extent that the frequency histograms in Figures 8.2a and 8.2b show differences between the distributions, those differences are not significant and the actual distribution can be described as resembling normal.

The third and most accurate way to make a judgement about criterion (2), is the Kolmogorov-Smirnov test. As noted for the one row chi-square test, if your data set does not resemble a normal distribution, the result of the test will be a $p$ value less than your stated significance level. If your data set does resemble a normal distribution, the result of the test will be greater than the stated significance level. However, given this test is best executed by specialist software, it is probably sufficient at undergraduate level to merely use the eyeball test, that is, to plot a frequency histogram, and compare it visually with a normal distribution and/or use the one row chi-square test.

Criterion (3) is often cited as the variance should be similar, since the standard deviation is derived from variance. They are both measures of spread; standard deviation is the square route of the variance. If it is asserted that an IV influences a DV, changes in the IV will merely shift the scores in the DV distribution along the horizontal scale. The effect of moving along the scale is to change the mean of the DV, not to disrupt its spread, standard deviation or variance. If the move of the distributions is perfect the ratio of the two standard deviations will be 1. The ratio is the larger number standard deviation over the smaller. The $F$ test is used to compare the ratio of the actual variances (or standard deviations) with critical values of $F$ in a table. If the calculated ratio is smaller than the value in the table, it is deemed that any differences between the standard deviations of the (two) variables is not significant. If the

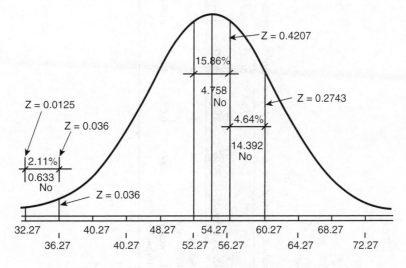

**Figure 8.1** Theoretical frequency counts for a normal distribution, *n*.

**Table 8.16** Actual frequency counts and theoretical frequency counts for a normal distribution, $n = 30$. Based upon data in Table 8.15.

| Class interval | Below 32.27 | 32.27 to 36.27 | 36.27 to 40.27 | 40.27 to 44.27 | 44.27 to 48.27 | 48.27 to 52.27 | 52.27 to 56.27 | 56.27 to 60.27 | 60.27 to 64.27 | 64.27 to 68.27 | 68.27 to 72.27 | 72.27 to 76.27 | Above 76.27 |
|---|---|---|---|---|---|---|---|---|---|---|---|---|---|
| Actual frequency | 0 | 1 | 2 | 1 | 5 | 2 | 9 | 3 | 3 | 0 | 3 | 1 | 0 |
| Expected frequency | 0.375 | 0.633 | 1.284 | 2.394 | 3.543 | 4.392 | 4.758 | 4.392 | 3.543 | 2.394 | 1.284 | 0.633 | 0.375 |

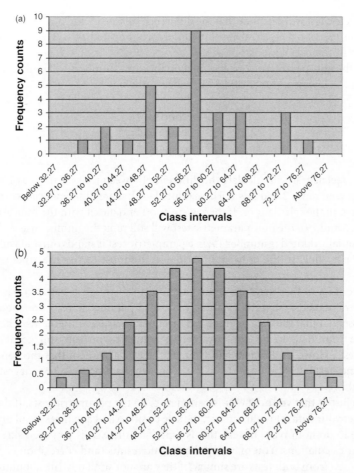

**Figure 8.2** Histograms of (a) actual and (b) theoretical frequencies for the data set in Table 8.15.

calculated ratio is higher than the value in the table, differences are significant, and the data set is deemed not parametric. Reading forward to the example in Table 8.26, the standard deviations for the two variables are 4.88 and 5.52. The ratio is $5.52/4.88 = 1.13$. The value of $n$ for both variables $= 30$ and therefore the $df = 29$. The critical value of $F$ from Appendix N is 1.85; therefore the standard deviations can be deemed similar, and against criterion (3), the data set is deemed parametric.

Parametric tests are more powerful and robust than non-parametric tests, and therefore more desirable. Parametric tests find differences in the data that non-parametric tests may not find. This is because the calculations underpinning the parametric tests are based on the absolute differences between sets of scores. This is not the case for the non-parametric tests; differences in scores are merely ranked. To illustrate the difference between how parametric and non-parametric tests manipulate data, take for example, three scores, thus: 3, 5 and 19.

The parametric test will calculate that the difference between 3 and 5 as 2, and the difference between 5 and 19 as 14. These real absolute differences of 2 and 14 will be carried forward to the next step in the calculations.

However, the non-parametric test executed on the same numbers will rank in sequence 3, 5 and 19, with the lowest number ranked number 1, thus:

3   becomes   1
5   becomes   2
19 becomes   3

The non-parametric test will calculate the difference between ranks 1 and 2 as 1, and also the difference between ranks 2 and 3 as 1. These diluted rank score differences are carried forward to the next step in the calculations. If there were another data set with the numbers 3, 5 and 6 (instead of 3, 5 and 19), the non-parametric test will still rank the number 6 as 3. The number that is substantially diluted is number 19; in a parametric test it stands out and will make a real impact on the $p$ value. In the non-parametric test, it almost carries the same weight as the number 6.

There is conflict in the literature about the need to distinguish between parametric and non-parametric data. Some authors argue that since parametric tests are robust, it is not absolutely essential that the data being tested meet parametric requirements. Therefore, even if you judge your data to be non-parametric, arguably you can still go ahead and do parametric tests. However, Bryman and Cramer's (1997) argument that the non-parametric Mann–Whitney test is 95% as powerful as the equivalent parametric '$t$' test suggests there is no need to do this.

If you can justify that your data set is parametric, do the parametric tests. If there is doubt about whether your data set is parametric, the safe position and as a measure of prudence may be to choose to do both types of test; this is especially the case if you are using software to perform your calculations. You may find the $p$ value results (and correlation coefficients for correlation tests) from both tests are similar; if they are not similar, write about this openly in your dissertation, and suggest how this impacts on your findings and conclusions. If you are clear your data are non-parametric, do the non-parametric tests. However, to demonstrate your understanding, you may also do the parametric tests, on the basis that some authors argue that it is not absolutely essential that the data being tested meet parametric require-ments. Again, if the $p$ values are not similar for both non-parametric and parametric tests, write about this in your findings and conclusions.

### Which difference in means test?

There are lots of choices of test. The first choice is to decide whether you are to undertake: (a) chi-square, or (b) difference in means, or (c) correlation. If you are clear you fall within (a) or (c), the options are few. The only choice in the chi-square test is the size of the contingency table, e.g. $2 \times 2$, or $3 \times 2$, etc. The only choices in correlation tests are the Spearman test for non-parametric data, or the Pearson test for parametric data. In difference in means tests the choices are many. When you select your tests, the issue is complicated in the statistical textbooks, since the same tests are sometimes referred to by many different names.

There are Internet pages that can be used to help in selecting the correct test; they can be located through search engines using key words such as 'which statistical test'. Perhaps it is better than you select logically in your own mind which test is most appropriate. The lead issues are:

- Be absolutely clear about your IV and DV
- Be clear about the level of measurement of both variables (categorical/nominal, ordinal, interval/ratio). Interval and ratio can be considered as though the same level of data when selecting the test

Given three possible levels of measurement for each variable, at stage one in Table 8.17, consider six possible permutations of test type. The table shows that for rows 1, 4, 5 and 6, you can go immediately to the appropriate test; only one stage of consideration is necessary.

For the difference in means tests, rows 2 and 3, a second stage of consideration is needed. There are three factors that drive the choice of difference in means test: (a) how many groups — classify as either two or more than two; (b) are the data sets unrelated or related; and (c) are the data sets non-parametric or parametric. It is absolutely essential that you are clear about (a) and (b). It is desirable that you are clear about whether your data sets are non-parametric or parametric (c). There are two options in each of these three factors, thus:

(a)  How many groups: classify as either two (A1) or more than two (A2) - more than two groups can also be labelled, perhaps more clearly, as three or more groups.
(b)  Are the data sets unrelated (B1) or related (B2).
(c)  Are the data sets non-parametric (C1) or parametric (C2).

Three factors, considered alongside each other with two values for each factor, give eight possible permutations as detailed in Table 8.18, and eight types of test. There are many other tests too for work at advanced levels. It is likely that the most often used test is the Mann–Whitney. At undergraduate level, it is probably perfectly acceptable, if your study is 'grouped' based, that you restrict yourself to just two groups. This will lessen the need to consider rows 5 to 8 in Table 8.18. You can substantiate just two groups in the context that your objectives should be modest, and there is no need, for example, to measure the sustainability performance (the DV) of construction materials in three groups, e.g. steel, concrete and timber, when better depth is possible and better knowledge may be gained by just examining two. If the articulation of your problem really forces one of your variables to be split into three or more groups, this is fine. However, you can still lessen the need to do tests identified in rows 5 to 8 of Table 8.18, by doing analysis in pairs, e.g. three groups A, B and C, all with a list of scores on some concept; perhaps sustainability performance. First test scores of group A versus group B, then group A versus group C, then group B versus group C. Whilst doing it this way means there are three tests instead of one, the output for a singular test, e.g. the Kruskal–Wallis test, will tell you if there is a significant difference somewhere in the three groups. It may be that two groups are similar, and one group is different from one or both of the others. Such an outcome will warrant further interrogation of the data, to determine precisely where differences lie; advanced studies may use Tukey post-hoc analysis.

**Table 8.17** Stage 1 in selecting which statistical test; based upon the level of measurement for the IV and DV.

| One variable (IV or DV) level of measurement | The other variable (IV or DV) level of measurement | Which test category? | Which test; name? Eleven choices |
|---|---|---|---|
| 1 Categorical/ nominal | Categorical/nominal | Chi-square | (1) Pearson's chi-square |
| 2 Categorical/ nominal | Ordinal | Difference in means (possibly chi-square if few values for the variable measured at the ordinal level) | Go to stage 2, Table 8.18, to select either: (2) Mann–Whitney (3) Wilcoxon (4) Friedman (5) Kruskal–Wallis |
| 3 Categorical/ nominal | Interval/ratio | Difference in means | Go to stage 2, Table 8.18, to select from tests (2) to (5), or additionally: (6) Unrelated '$t$' test (7) Related '$t$' test (8) Unrelated ANOVA (9) Related ANOVA |
| 4 Ordinal | Ordinal | Correlation | (10) Spearman's rho |
| 5 Ordinal | Interval/ratio | Correlation | |
| 6 Interval/ratio | Interval/ratio | Correlation | (11) Pearson's product moment |

In the same sort of way that language used by statisticians can be unhelpful when different words or phrases are used to describe unrelated and related data (Table 8.14), the language is also unhelpful when naming tests. For example, the Mann–Whitney $U$-test is also called a '$t$' test (a non-parametric unrelated version of a '$t$' test). Also the Wilcoxon is also called a '$T$'

**Table 8.18** Stage 2 in selecting which difference in means test.

| How many groups; two (A1) or three or more (A2) | Are the data sets unrelated (B1) or related (B2) | Are the data sets non-parametric (C1) or parametric (C2) | Type of test |
|---|---|---|---|
| 1 Two groups (A1) | Unrelated (B1) | Non-parametric (C1) | Mann–Whitney $U$ test |
| 2 Two groups (A1) | Unrelated (B1) | Parametric (C2) | Unrelated '$t$' test |
| 3 Two groups (A1) | Related (B2) | Non-parametric (C1) | Wilcoxon matched pairs signed rank |
| 4 Two groups (A1) | Related (B2) | Parametric (C2) | Related '$t$' test |
| 5 Three or more groups (A2) | Unrelated (B1) | Non-parametric (C1) | Kruskal–Wallis |
| 6 Three or more groups (A2) | Unrelated (B1) | Parametric (C2) | Unrelated ANOVA |
| 7 Three or more groups (A2) | Related (B2) | Non-parametric (C1) | Friedman |
| 8 Three or more groups (A2) | Related (B2) | Parametric (C2) | Related ANOVA |

test, but with a capital letter (a non-parametric related version). The unrelated '*t*' test (row 2), is also called, amongst other things, an independent '*t*' test. An ANOVA, the name used for some tests involving three or more groups, is merely an abbreviation for 'analysis of variance'. This name can be used to loosely describe all inferential tests, since whether there are two or more groups, providing there are at least two variables, all tests are analysing variance. You need to try to cut your way through the language, as far as is possible.

The tests compare mean scores and distributions around the means of two (or more) sets of numbers; that is, a set of numbers that are in two separate groups. When the two sets of numbers are compared, if their means and distributions are 'significantly' different, it is inferred that the two groups are indeed different and that they represent two different samples. When interpreting the tests, it is inferred that the IV influences the DV; the null hypothesis will be rejected. If the difference between the distributions is small, it is inferred that the two sets of numbers are from the same sample. There is a relationship between the variables; the null hypothesis cannot rejected. In this latter case, to the extent that there are differences between the distributions, these differences occur due to chance.

## 8.5  Difference in means: the unrelated Mann–Whitney test

To illustrate a difference in means test, the same scenario and study objective is used from the chi-square test in Table 8.1. The spreadsheet is adapted; the raw data set is now in Table 8.19. The change between the two scenarios is the level of measurement for the DV, cost predictability. For the chi-square test, cost predictability data were only available at the categorical/nominal level, in two groups, (a) on or below budget, and (b) over budget. For the difference in means test, cost predictability data are available at the interval level; that is, the percentage figure above or below budget. Since this is a hypothetical example, for the chi-square test, assume data for the DV were not available at the interval level; if it had been available, the data would have been used at the interval level, and the difference in mean test would be undertaken in preference to the chi-square test.

Repeating important data used in the chi-square test: the objective remains to 'determine whether method of procuring work influences cost predictability'. As a null hypothesis for testing, this is written 'method of procuring work does not influence cost predictability'. The significance level is set at $p \leq 0.05$. The IV is method of procuring work. Method of procuring work is in two 'groups' or it has two 'values': (a) competitive bidding, coded number 1, and (b) collaborative working, coded number 2. The DV is cost predictability measured on completion of projects. It is measured at the interval level in terms of percentages. Data are obtained from thirty completed projects, selected randomly.

### Assembling the raw data

Use a spreadsheet to assemble and sort the raw data; number each of the projects 1 to 30 in column 1. In column 2, label each project's method of procurement. In column 3, label each project's percentage below or above budget. Use columns 4 and 5 to split the IV into two groups; column 3 is values of the DV for projects procured competitively, and column 4 is

**Table 8.19** Raw data for Mann–Whitney difference in means test.

| 1 Project number | 2 The IV: method of procurement. Competitive bidding = 1, collaborative working = 2 | 3 The DV: cost predictability. Percentage change on budget | 4 Percentage change on budget for competitive bidding projects (group A) | 5 Percentage change on budget for collaborative working projects (group B) |
|---|---|---|---|---|
| 1 | 2 | +5.0 | | +5.0 |
| 2 | 2 | 0 | | 0 |
| 3 | 1 | 0 | 0 | |
| 4 | 1 | −1.1 | −1.1 | |
| 5 | 1 | +15.3 | +15.3 | |
| 6 | 2 | +2.5 | | +2.5 |
| 7 | 1 | +1.0 | +1.0 | |
| 8 | 1 | +8.3 | +8.3 | |
| 9 | 1 | +6.2 | +6.2 | |
| 10 | 2 | +4.8 | | +4.8 |
| 11 | 2 | −3.0 | | −3.0 |
| 12 | 1 | +4.5 | +4.5 | |
| 13 | 2 | −2.8 | | −2.8 |
| 14 | 1 | −7.2 | −7.2 | |
| 15 | 2 | −1.0 | | −1.0 |
| 16 | 1 | 0 | 0 | |
| 17 | 2 | −0.1 | | −0.1 |
| 18 | 1 | 0 | 0 | |
| 19 | 1 | +6.6 | +6.6 | |
| 20 | 1 | −0.2 | −0.2 | |
| 21 | 2 | −2.2 | | −2.2 |
| 22 | 2 | −5.2 | | −5.2 |
| 23 | 1 | +8.2 | +8.2 | |
| 24 | 1 | +2.0 | +2.0 | |
| 25 | 1 | +1.2 | +1.2 | |
| 26 | 1 | +3.2 | +3.2 | |
| 27 | 1 | +10.0 | +10.0 | |
| 28 | 2 | −3.0 | | −3.0 |
| 29 | 2 | −2.2 | | −2.2 |
| 30 | 2 | −2.9 | | −2.9 |
| Total | | 47.9 | 58 | −10.1 |
| n | 1 = 17 No, 2 = 13 No | 30 | 17 | 13 |
| Mean score (total/n) | | 1.59 | +3.41 | −0.78 |

values of the DV for projects procured in collaboration. The mean scores, a key output from the test, are calculated at the bottom of the table. To emphasize their importance, they are separated here from the main text, thus:

- Mean cost predictability for competitive bid projects = +3.41%
- Mean cost predictability for collaboratively bid projects = −0.78%

As part of the stage 1 analysis, eyeballing the spread of the scores in Table 8.19 and more easily, eyeballing the mean scores, suggests that the IV may be influencing the DV; method of

**Table 8.20** Data arranged into class intervals of 4% points for a frequency histogram

| Percentage class intervals | Procured competitively | Procured collaboratively |
|---|---|---|
| −8.0 to −4.0 | 1 | 1 |
| −3.9 to 0 | 5 | 9 |
| +0.1 to +4.0 | 4 | 1 |
| +4.1 to +7.9 | 3 | 2 |
| +8.0 to +11.9 | 3 | 0 |
| +12.0 to +15.9 | 1 | 0 |
| Total | 17 | 13 |

procurement influences cost predictability. The data set shows that collaborative working projects complete, on average, 0.78% below budget. Competitively bid projects complete, on average, 3.41% over budget. A frequency histogram of the data can be drawn to support the eyeball observation. However, to produce the histogram in a meaningful format, the data need to be arranged into appropriate class intervals. In this case four percentage points is used, as illustrated in Table 8.20. The frequency histograms are shown in Figures 8.3a and 8.3b. It is readily apparent that, for competitively procured projects, some finish on or under budget, but most finish over budget. Also for projects procured collaboratively, some finish over budget, but the vast majority are on or under budget. Similar to the chi-square test, if you feel unable to go to stage 2 of the analysis, just do stage 1. Be sure to illustrate your understanding of the mean scores in stage 1.

### Are differences due to chance? Manual calculations: p values and degrees of freedom

Stage 2 determines whether, to the extent that there are differences between the distributions of the scores in Figures 8.3a and 8.3b, and the mean scores themselves, could those differences have occurred due to *chance*? There are two values for Mann–Whitney $U$ to be calculated. The

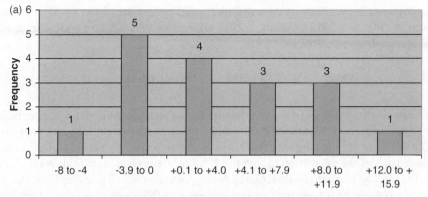

**Class intervals: cost predictability.  Percentage change on budget**

**Figure 8.3** Frequency histograms for the cost predictability of projects procured (a) competitively and (b) collaboratively.

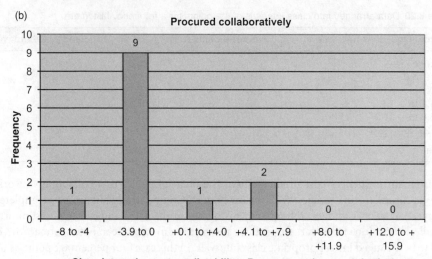

**Figure 8.3** (*Continued*)

larger number, which is known notionally as U′, pronounced 'U prime', is discarded. The smaller number, which is called U, is used for interpretation of the p value in the table. The U value is the penultimate figure in the calculation; it is used in the table in Appendix H to determine the p value. The p value will be used, alongside the mean scores and other descriptive statistics, in interpretation of the result. The first U is calculated from the mathematical formula, thus:

$$\text{Mann-Whitney } U = \frac{NaNb + Na(Na + 1)}{2} - \Sigma Ra$$

Where: Na is the number of scores in group 'A', Nb is the number of scores in group 'B', Ra is the sum of ranks in list 'A', smallest score rank 1, and $\Sigma$ is the Greek letter sigma, representing 'the sum of'. Mann–Whitney U is the critical value to be assessed for a p value in the table.

The second U is calculated by:

$$NaNb - \text{the result of the calculation from the first U}$$

It is the raw numbers that will drive whether the first or second calculation will give the lower value U. From observation of the first U formula, although there are three parts, there are just two key elements: (a) the number of scores in each group, and (b) the sum of the rankings. The three parts of the calculation to find the U value, are presented in Table 8.21. The table allows the ranking scores to be determined easily (see example in Table 7.6. Note that in the ranking procedure, the two separate groups are treated as though one (not ranked separately), and only the sum of the ranks for group 'A' are used. The fourth part of the Mann–Whitney calculation, is to determine the p value, by comparing the calculated value of U with the critical values of U in a table.

**Table 8.21** Computation of the Mann–Whitney $U$ value.

**Number of scores in group 'A' = 17, Number of scores in group 'B' = 13**

**Part 1: NaNb = 17 × 13 = 221**

**Part 2:** $\frac{Na(Na+1)}{2} = \frac{17(17+1)}{2} = 153$

**Part 3**

| | Group 'A' scores | Group 'A' ranking | Group 'B' scores | Group 'B' ranking |
|---|---|---|---|---|
| 1 | 0 | 14.5 | +5.0 | 24 |
| 2 | −1.1 | 9 | 0 | 14.5 |
| 3 | +15.3 | 29.5 | +2.5 | 20 |
| 4 | +1.0 | 29.5 | +4.8 | 23 |
| 5 | +8.3 | 28 | −3.0 | 3.5 |
| 6 | +6.2 | 25 | −2.8 | 6 |
| 7 | +4.5 | 22 | −1.0 | 10 |
| 8 | −7.2 | 1 | −0.1 | 12 |
| 9 | 0 | 14.5 | −2.2 | 7.5 |
| 10 | 0 | 14.5 | −5.2 | 2 |
| 11 | +6.6 | 26 | −3.0 | 3.5 |
| 12 | −0.2 | 11 | −2.2 | 7.5 |
| 13 | +8.2 | 27 | −2.9 | 5 |
| 14 | +2.0 | 19 | | |
| 15 | +1.2 | 18 | | |
| 16 | +3.2 | 21 | | |
| 17 | +10.0 | 2 | | |
| Total | 58 | 311.5 (ΣRa) | 10.1 | |
| Total '$n$' | 17 | n.a. | 13 | |
| Mean (total score/$n$) | +3.41 | n.a | −0.78 | |

Mann–Whitney $U = NaNb + \frac{Na(Na+1)}{2} - \Sigma Ra$

Mann–Whitney $U =$ part 1 + part 2 − part 3

The first Mann–Whitney $U = (221 + 153) - 311.5 = 62.5$

The second Mann–Whitney $U = NaNb -$ the result of the calculation from the first
$U = (17 \times 13) - 62.5 = 158.5$

$U'$ the larger figure, discarded = 158.5. $U$ taken to the table = 62.5. The value in the table, for group 'A' $n$ of 17 and group 'B' $n$ of 13, is 63

If there are small differences between mean scores and the spread of numbers, the $p$ value will be higher than the set significance level; the finding will be 'cannot reject the null hypothesis, the IV does not influence the DV'. Alternatively, if there are large differences between mean scores and the spread of numbers, the $p$ value will be lower than the set significance level; the finding will be 'reject the null hypothesis, the IV does influence the DV'.

For the stated significance level of $p \leq 0.05$, for group 'A' $n$ of 17 and group 'B' $n$ of 13, the critical $U$ value in the table in Appendix H is 63. Our value of 62.5 does not exceed the critical value, and we can reject the null hypothesis with the significance level set at $\leq 0.05$. The finding is: reject the null hypothesis with $p$ set at $\leq 0.05$; method of procurement influences cost predictability. The mean scores are competitively bid projects $+3.41\%$, collaborative working projects $= -0.78\%$. Collaborative working projects are more likely than competitively bid projects to complete projects on or below budget.

It is notable that using the same scenario and base data set, the chi-square test does not find a significant difference, but the Mann–Whitney test does. That is because the Mann–Whitney test is working with richer data for the DV (cost predictability is at the interval level), whereas for the chi-square test the data set for cost predictability is only available at the categorical/nominal level.

### The Consequence of larger sample size and different spread of numbers

Consider again the data in Table 8.19. Suppose data were available for a further 30 projects, such that $n = 60$. Suppose instead of 17 competitively procured projects and 13 collaboratively procured projects, there were now data for 34 and 26 projects respectively. Suppose hypothetically, by some fluke, the data set for the further projects were identical to the original data set. This would mean that mean scores, standard deviation and other descriptive statistics are the same. Since the sample size is higher, the $p$ value for the Mann–Whitney test, calculated by Excel, for $n = 34$ group 'A' and $n = 26$ group 'B' $= 0.0002$, but the $p$ value for $n = 16$ group 'A' and $n = 13$ group 'B' $= 0.012$. Therefore, with a similar distribution of scores, but with higher sample size, we can be more confident that differences that may occur are not due to chance.

Still considering the data in Table 8.19; suppose $n$ remains at 30 and the mean scores remain at 3.41% and $-0.78\%$. However, the distribution and spread of the numbers is changed by adding and deducting the arbitrary score of 10 thus:

- Of the 17 competitively procured projects, eight are increased by 10 points, eight are decreased by 10 points and one is left the same
- Of the 13 projects procured collaboratively, six are increased by 10 points, six are decreased by 10 points and one is left the same

The calculated $p$ value changes from 0.012 to 0.339. This large change to the $p$ value, as a consequence of a change to the spread of the numbers, without a change in the difference to the means, illustrates that it is really a simplification to describe the tests as merely 'difference in means'.

## 8.6 Difference in means: the related Wilcoxon test

To illustrate a related difference in means test, a similar scenario is used from the chi-square test and the Mann–Whitney. The DV remains cost predictability, but the articulation of the problem is not around method of procurement, but around the cost predictability of estimates. Therefore, the whole basis and design of the study changes. Cost estimates are available for two stages in the process: (a) at client instruction to proceed, and (b) value of order placed with contractor, including contingency. The final account figures are given, and percentage changes of estimates benchmarked against these figures.

**Table 8.22** Raw data for Wilcoxon difference in means test.

| 1 Project number | 2 Estimate at pre-tender £ | 3 Estimate at pre-start £ | 4 Final account £ | 5 Percentage change on pre-tender budget | 6 Percentage change on pre-start budget |
|---|---|---|---|---|---|
| 1 | 121 000 | 125 100 | 131 355 | 8.56 | + 5.0 |
| 2 | 80 000 | 100 080 | 100 080 | 25.1 | 0 |
| 3 | 852 00 | 950 220 | 950 220 | 11.53 | 0 |
| 4 | 650 000 | 610 025 | 603 315 | −7.18 | −1.1 |
| 5 | 623 000 | 750 123 | 864 892 | 38.83 | + 15.3 |
| 6 | 85 000 | 85 789 | 87 934 | 3.45 | + 2.5 |
| 7 | 900 000 | 850 456 | 858 961 | −4.56 | + 1.0 |
| 8 | 750 000 | 900 120 | 974 830 | 29.98 | + 8.3 |
| 9 | 450 000 | 525 00 | 557 658 | 23.92 | + 6.2 |
| 10 | 424 000 | 450 458 | 472 080 | 11.34 | + 4.8 |
| 11 | 363 000 | 381 235 | 369 798 | 1.87 | −3.0 |
| 12 | 190 000 | 195 785 | 204 595 | 7.68 | + 4.5 |
| 13 | 220 000 | 230 482 | 224 029 | 1.83 | −2.8 |
| 14 | 565 000 | 575 585 | 534 124 | −5.46 | −7.2 |
| 15 | 565 000 | 565 220 | 559 568 | −0.96 | −1.0 |
| 16 | 575 000 | 590 589 | 590 569 | 2.71 | 0 |
| 17 | 100 000 | 100 981 | 99 971 | −0.03 | −0.1 |
| 18 | 90 000 | 80 222 | 80 222 | − 10.86 | 0 |
| 19 | 650 000 | 650 666 | 693 610 | 6.71 | + 6.6 |
| 20 | 120 000 | 130 123 | 129 863 | 8.22 | −0.2 |
| 21 | 195 000 | 250 555 | 245 043 | 25.66 | −2.2 |
| 22 | 200 000 | 190 855 | 180 931 | −9.53 | −5.2 |
| 23 | 800 000 | 800 780 | 866 444 | 8.31 | + 8.2 |
| 24 | 825 000 | 850 335 | 867 342 | 5.13 | + 2.0 |
| 25 | 440 000 | 450 665 | 456 073 | 3.65 | + 1.2 |
| 26 | 725 000 | 800 500 | 826 116 | 13.95 | + 3.2 |
| 27 | 890 000 | 950 222 | 1 045 244 | 17.44 | + 10.0 |
| 28 | 125 000 | 135 110 | 131 057 | 4.85 | −3.0 |
| 29 | 510 000 | 525 000 | 513 450 | 0.68 | −2.2 |
| 30 | 555 000 | 560 010 | 543 770 | −2.02 | −2.9 |
| Total | | | | 220.77 | 47.90 |
| Mean | | | | 7.36 | 1.60 |

Table 8.22, adapted from Table 8.19, illustrates the raw data set. The objective is to 'compare cost predictability of estimates at pre-tender and pre-start stages'. As a null hypothesis for testing, this is written 'stage of estimate does not influence cost predictability'. The significance level is set at $p \leq 0.05$. The IV is stage of estimate; it is in two 'groups' or it has two 'values': (a) pre-tender, coded number 1, and (b) pre-start, coded number 2. The DV is cost predictability measured on completion of projects. It is measured at the interval level in terms of percentages. Data are obtained from 30 completed projects, selected randomly.

Table 8.23 represents columns 5 and 6 of the raw data set in Table 8.22, as new columns 1 and 2. Column 3 is the numerical difference of column 1 minus column 2, with negative differences noted. Columns 4 and 5 record the differences; positive differences in column 4 and negative differences in column 5. The ranking of data from column 3 is carried out

**Table 8.23** Wilcoxon difference in means test.

| | 1 Pre-tender percentage change | 2 Pre-start percentage change | 3 Difference | 4 Ranks for positive differences | 5 Ranks for negative differences |
|---|---|---|---|---|---|
| 1 | 8.56 | +5.0 | 3.56 | 12 | |
| 2 | 25.1 | 0 | 25.10 | 28 | |
| 3 | 11.53 | 0 | 11.53 | 24 | |
| 4 | −7.18 | −1.1 | −6.08 | | 17 |
| 5 | 38.83 | +15.3 | 23.53 | 27 | |
| 6 | 3.45 | +2.5 | 0.95 | 6 | |
| 7 | −4.56 | +1.0 | −5.56 | | 16 |
| 8 | 29.98 | +8.3 | 21.68 | 26 | |
| 9 | 23.92 | +6.2 | 17.72 | 25 | |
| 10 | 11.34 | +4.8 | 6.54 | 18 | |
| 11 | 1.87 | −3.0 | 4.87 | 15 | |
| 12 | 7.68 | +4.5 | 3.18 | 11 | |
| 13 | 1.83 | −2.8 | 4.63 | 14 | |
| 14 | −5.46 | −7.2 | 1.74 | 7 | |
| 15 | −0.96 | −1.0 | 0.04 | 1 | |
| 16 | 2.71 | 0 | 2.71 | 29 | |
| 17 | −0.03 | −0.1 | 0.07 | 2 | |
| 18 | −10.86 | 0 | −10.86 | | 23 |
| 19 | 6.71 | +6.6 | 0.11 | 3 | |
| 20 | 8.22 | −0.2 | 8.42 | 21 | |
| 21 | 25.66 | −2.2 | 27.86 | 30 | |
| 22 | −9.53 | −5.2 | −4.33 | | 13 |
| 23 | 8.31 | +8.2 | 0.11 | 3 | |
| 24 | 5.13 | +2.0 | 3.13 | 10 | |
| 25 | 3.65 | +1.2 | 2.45 | 8 | |
| 26 | 13.95 | +3.2 | 10.75 | 22 | |
| 27 | 17.44 | +10.0 | 7.44 | 19 | |
| 28 | 4.85 | −3.0 | 7.85 | 20 | |
| 29 | 0.68 | −2.2 | 2.88 | 9 | |
| 30 | −2.02 | −2.9 | 0.88 | 5 | |
| Total | 220.77 | 47.90 | | 395 | 69 |
| Mean | 7.36 | 1.60 | | | |

ignoring negative and positive signs. Columns 5 and 6 are totalled. The smaller of the two totals for columns 5 and 6 is the value of $t$. Our values are 395 and 69; the value of $t$ is 69.

For the stated significance level of $p \leq 0.05$, the critical $t$ value in the table in Appendix J for $n = 30$ is 137. The calculated value is required to be equal to or less than the stated value to be significant. Our value of 69 is less than the critical value, and we can reject the null hypothesis with the significance level set at $\leq 0.05$. Our value of 69 is also less than the critical value for $p \leq 0.01$, which from the table is 109. The finding is: reject the null hypothesis with $p$ set at $\leq 0.05$; 'stage of estimate influences cost predictability'. Estimates at pre-start stage (1.6% under-estimate) are significantly more accurate than estimates at pre-tender stage (7.36% under-estimate). The finding is incidentally also significant at the value of $p \leq 0.01$. Using computer software, the calculated $p$ value is 0.002.

## 8.7 Difference in means: the parametric related *t* test

The data set in Table 8.23 is represented in columns (i) and (ii) of Table 8.24, with columns (iii) and (iv) added to compute the related *t* test. This parametric related *t* test is exactly the same as the non-parametric related Wilcoxon test. Using the related *t* test instead of the Wilcoxon test relies on the assumption that the data set is parametric. For the purposes of illustrating the related *t* test calculation, this assumption is taken here. A key issue to note in the parametric tests is that the computation requires the calculation of real differences between numbers, whereas the non-parametric tests require only ranking. Column (iii) is the difference between columns (i) and (ii). Do not be put off by the formula for the related *t* test, which is thus:

$$t = \frac{|\bar{X} - \bar{Y}|}{\sqrt{\frac{\Sigma D^2 - \frac{(\Sigma D)^2}{N}}{N(N-1)}}}$$

Where: $\bar{X}$ = the mean of the first list, $\bar{Y}$ = the mean of the second list, $D$ = the difference between paired $\bar{X}$ bar and $\bar{Y}$ bar scores, $N$ = the number of score, $\Sigma D^2$ = the differences squared, then totalled and $(\Sigma D)^2$ = the differences totalled, then squared.

To simplify the calculation, it is undertaken in eight parts; for ease the first three parts are presented in Table 8.24

Part 1 = $\bar{X} - \bar{Y}$ = 5.76
Part 2 = $\Sigma D^2$ = 3531
Part 3 = $(\Sigma D)^2/N$ = 996
Part 4 = $N(N-1)$ = 30 (30 − 1) = 30 × 29 = 870
Part 5 = part 2 − part 3/part 4 = 3531 − 996/870 = 2535/870 = 2.91
Part 6 = square root of part 5 = $\sqrt{2.91}$ = 1.70
Part 7; the critical value of *t* = part 1/part 6 = 5.76/1.70 = 3.38
Part 8; observe the stated critical value of *t*, from the table. Calculated value of *t* must be equal to or more than the stated value to be significant. For the significance level of $p \leq 0.05$, the critical *t* value for degrees of freedom of 29 $(n-1)$ = 2.045

Our value of 3.38 exceeds the critical value, and we can reject the null hypothesis with the significance level set at $\leq$ 0.05. Our value of 3.38 is also more than the critical value for $p \leq 0.01$, which from the table is 2.75. The finding is: reject the null hypothesis with *p* set at $\leq$ 0.05; 'stage of estimate influences cost predictability'. Estimates at pre-start stage (1.6% under-estimate) are significantly more accurate than estimates at pre-tender stage (7.36% under-estimate). The finding is incidentally also significant at the value of $p \leq 0.01$.

## 8.8 Correlations

Correlations are richer than difference in mean and chi-square tests. The correlation coefficients, as a key output from the tests, are more informative than mean scores or contingency tables. To use correlation tests, both of the variables need to be at the ordinal or

**Table 8.24** Computation of the parametric related *t* test.

| | Part 1 $\bar{X}-\bar{Y}$ | | Part 3 | Part 2 |
|---|---|---|---|---|
| | Column (i); $\bar{X}$ | Column (ii); $\bar{Y}$ | Column (iii); D | Column (iv); $D^2$ |
| 1 | 8.56 | +5.0 | 3.56 | 12.66 |
| 2 | 25.1 | 0 | 25.10 | 630.01 |
| 3 | 11.53 | 0 | 11.53 | 132.90 |
| 4 | −7.18 | −1.1 | −6.08 | 36.99 |
| 5 | 38.83 | +15.3 | 23.53 | 553.52 |
| 6 | 3.45 | +2.5 | 0.95 | 0.91 |
| 7 | −4.56 | +1.0 | −5.56 | 30.91 |
| 8 | 29.98 | +8.3 | 21.68 | 469.91 |
| 9 | 23.92 | +6.2 | 17.72 | 314.13 |
| 10 | 11.34 | +4.8 | 6.54 | 42.77 |
| 11 | 1.87 | −3.0 | 4.87 | 23.74 |
| 12 | 7.68 | +4.5 | 3.18 | 10.12 |
| 13 | 1.83 | −2.8 | 4.63 | 21.45 |
| 14 | −5.46 | −7.2 | 1.74 | 3.01 |
| 15 | −0.96 | −1.0 | −0.04 | 0.00 |
| 16 | 2.71 | 0 | 2.71 | 7.33 |
| 17 | −0.03 | −0.1 | 0.07 | 0.01 |
| 18 | −10.86 | 0 | −10.86 | 118.04 |
| 19 | 6.71 | +6.6 | 0.11 | 0.01 |
| 20 | 8.22 | −0.2 | 8.42 | 70.88 |
| 21 | 25.66 | −2.2 | 27.86 | 776.35 |
| 22 | −9.53 | −5.2 | −4.33 | 18.79 |
| 23 | 8.31 | +8.2 | 0.11 | 0.01 |
| 24 | 5.13 | +2.0 | 3.13 | 9.81 |
| 25 | 3.65 | +1.2 | 2.45 | 6.02 |
| 26 | 13.95 | +3.2 | 10.75 | 115.50 |
| 27 | 17.44 | +10.0 | 7.44 | 55.40 |
| 28 | 4.85 | −3.0 | 7.85 | 61.55 |
| 29 | 0.68 | −2.2 | 2.88 | 8.28 |
| 30 | −2.02 | −2.9 | 0.88 | 0.77 |
| Total | 220.77 | 47.90 | $\Sigma D = 172.87$ | 3531.75 say 3531 |
| mean | 7.36 | 1.60 | $(\Sigma D)^2 = 29884$ | |
| | $\bar{X}-\bar{Y} = 7.36 - 1.60 = 5.76$ | | $(\Sigma D)^2/N = 29884/30 = 996$ | |

interval/ratio level. Scores are paired; a score on one variable is automatically paired with a score on another variable. The issue of whether the data sets are non-parametric or parametric needs to be considered for both variables. Both need to be parametric to use the Pearson correlation. If one or both are non-parametric, the Spearman Rho test is appropriate.

The IV and the DV may be measures of variables using the same units, e.g. interest rates and inflation rates both expressed as percentages. Alternatively, they may have completely different measurement units, e.g. cement content in concrete kg/m$^3$, and compressive strength of concrete N/mm$^2$. Measures such as interest rates and inflation again (there are lots of others too) may be paired by time periods, e.g. each calendar month over several years, or average annual rates paired over many years. The two variables often measure different concepts; they may, however, measure exactly the same concept, but perhaps in two different countries, e.g. construction unemployment figures over a given period.

There is the possibility to take linear correlations to the next stage of analysis; linear regression. This involves calculating a line of best fit and plotting it on the scatter graph. This will allow, given the score of one variable, a prediction of a score of the other variable. If you wish to do this, refer to a specialist statistical text such as Hinton (2004).

As a precursor to starting the correlation tests, as always, you should undertake the descriptive analysis around both variables; that is, range, maximum, minimum, mean, median, mode, and standard deviation. There is a lot of potential for discussion around the mean scores in their own right; are they higher or lower than ought reasonably be expected?

Two stages in the analysis can be identified: (1) eyeball the scatter diagram, and (2) calculate a $p$ value, correlation coefficient and correlation coefficient squared. As with the chi-square and difference in mean tests, if you feel unable to carry out stage 2 of the analysis, just do stage 1. The first stage of eyeballing data in a scatter diagram is extremely important. There are several issues to look out for. The analogy of an egg-shaped rugby ball may be useful:

- The data may cluster around a theoretical straight line, as though close to a narrow rugby ball; this indicates a strong relationship. A given manipulation of the IV may result in a strong movement of the DV
- The data may cluster around a theoretical straight line, as though a fat rugby ball; this indicates a relationship, though not so strong. A given manipulation of the IV may result in a modest movement of the DV
- The rugby ball, or the theoretical line through it, may go up left to right indicating a positive relationship. A positive relationship occurs if manipulating the IV up impacts on the DV similarly moving up (or manipulating the IV down, impacts on the DV similarly moving down), e.g. as cement content in concrete increases, compressive strength of concrete increases
- The rugby ball, or the theoretical line through it, may go up right to left indicating a negative relationship. A negative relationship occurs if manipulating the IV up impacts on the DV moving down (or manipulating the IV down, impacts on the DV moving up), e.g. as interest rates move up, convention in economics is that inflation rates will move down
- The shape and pattern of the data set may not cluster; it may be completely random or square or close to circular. This indicates no relationship
- The data may be parabolic or curve linear; this would require careful interpretation
- There may be outliers; the general spread may indicate a relationship, but there may be one or two measures. Consideration may be given to whether such measures are spurious; they may be just bad or faulty data. Analysis and findings can be undertaken with and without the outliers present in the data set
- The relationship between the variables may not be consistent along the whole length of a theoretical straight line; the scatter plot may indicate clusters of data that do not sit well with the rest of the data (the $t$ test in Excel calls this 'homoscedastic')

Be mindful that, in some cases, the direction of the correlation coefficient, positive or negative, has no real meaning. Some variables, measured on an interval scale, will be arbitrarily anchored as high or low, e.g. in some studies, McGregor's theory X theory Y, is arbitrarily anchored with Y as the high score, but it could equally be anchored with X as high.

## *Are differences due to chance?; and the correlation coefficient*

The second stage of the analysis is to calculate a $p$ value, correlation coefficient and correlation coefficient squared. Having completed the first stage eyeballing, with practice, you may be interested, before undertaking the calculations, to estimate what the $p$ value and correlation coefficients might be at the end. If the $p$ value is less than or equal to the set significance level (e.g. $p \leq 0.05$), it is then appropriate to consider the correlation coefficient, since any relationship indicated by the subsequent correlation coefficient has not occurred due to chance. If the $p$ value is more than the set significance level, the subsequent correlation coefficient is irrelevant, since to the extent that any relationship may be apparent, it may have occurred due to chance.

$P$ values always lie in the range of 0.00 to $+1.00$. Correlation coefficients always lie in the range of $-1.00$ to $+1.00$. The two figures must be reported, and must not be confused; their meanings are fundamentally different. Strong correlations are close to $-1$ or $+1$, and they resemble a straight line relationship. A correlation coefficient of 1.0 or $-1.0$ is perfect; correlation coefficients of 1.0 or close to 1.0 rarely occur. If a strong relationship is found, this may give a correlation coefficient of say 0.80 or $-0.80$, and a $p$ value of 0.05; the scatter graph may appear as though it could be surrounded by a thin rugby ball, e.g. Figure 8.5a. A correlation of $-0.50$ or $+0.50$ tends to be replicate a fatter rugby ball, e.g. Figure 8.5c. Studies may report success, with correlation coefficients of circa 0.30 or $-0.30$. A correlation close to zero, on a scatter plot, looks as though the points could be encircled by a football. Though a correlation coefficient is a numerical value on a wide scale, the process of managing information is such that qualitative data are converted to quantitative, and vice versa; therefore qualitative labels are loosely ascribed to the quantitative correlation coefficients, e.g.:

- To 0.25 (or $-0.25$) = little or no relationship
- 0.26 to 0.50 = fair degree of relationship
- 0.51 to 0.75 = moderate to good relationship
- 0.75 to 1.00 = very good to excellent relationship

Alternatively, Cohen and Holliday (1996, pp.82–83) offer their 'rough and ready guide':

- To 0.19 = a very low correlation
- 0.20 to 0.39 = a low correlation
- 0.40 to 0.69 = a modest correlation
- 0.70 to 0.89 = a high correlation
- 0.90 to 1.00 a very high correlation

Alongside the correlation coefficient, you should also calculate the correlation coefficient squared (or r squared), sometimes called the coefficient of determination; it always lies between 0 and $+1.00$. R squared indicates the 'percentage of the variation in one variable that is related to the variation in the other'. A correlation coefficient of 0.90 or $-0.90$ thus becomes $+0.81$, or 0.50 becomes 0.25; in interpretation they are then expressed as a percentage, e.g. 25%, meaning that 25% of the movement in one variable (the DV) is due to movement in the other variable (the IV). This concept is useful in understanding that it is not only the IV that influences the DV; it sits very well alongside the phrase in Chapter 4 'there are lots of variables at large'. If the DV is for example profit, there are clearly lots of variables that influence this; if your study can just identify one variable that has a very modest influence on profit, this would be worthwhile. You should note that some statisticians argue it is not appropriate to compute $r^2$ from the non-parametric Spearman correlation coefficient.

When interpreting the test, if a *p* value is found equal to or less than the stated significance level, the null hypothesis will be rejected; there is a relationship between the variables. If a *p* value is found to be more than the stated significance level, the null hypothesis cannot be rejected; there is no relationship found between the variables.

### Manual calculations for Spearman's Rho

Spearman's rho (sometimes written $\rho$ or $r_s$) is calculated from the mathematical formula, thus:

$$\text{Spearman's } rho = 1 - \frac{6\Sigma D^2}{(N^3 - N)}$$

Where: Spearman's rho is the critical value to be assessed for a *p* value in the table, $\Sigma$ is 'the sum of', D is the differences between ranked scores and N is the number of paired scores.

To illustrate a correlation test, initially the same scenario and study objective is used from the chi-square test in Table 8.1 and is represented in Table 8.25. In Table 8.1, method of procurement is classified in two groups as competitive or collaborative. A third middle group label is identified; competitive dialogue. With insight on each project, it may be possible to subjectively score method of procurement on a scale of 0 to 10 thus:

- 0–1: extremely competitive, open and advertised lists
- 2–3: competitive; bidders from select list
- 4–6: some competition, with move towards collaboration, pre-qualification of bidders, criteria for selection include issues other than price
- 7–8: intent to collaborate, flavoured by strong presence of competition, few bidders
- 9–10: extremely collaborative, two bidders or single bidder, negotiated price or budget

Calculated using computer software, using the data in columns 2 and 3, the *p* value for these data is $\leq 0.05$, Pearson's correlation coefficient is $-0.437$, and $r^2$ is 0.19. The negative relationship arises because the high score on method of procurement is collaborative working, and this equates with a low scores on cost predictability. The scatter diagram, if drawn, would illustrate this clearly.

A new variable is introduced to change the theme for the manual calculation; that is, time predictability, as illustrated in Table 8.26. The objective is to 'determine whether there is a relationship between time and cost predictability of projects'. Time predictability is named as the IV, and cost predictability as the DV; it makes for interesting debate whether it could be the other way round. The objective emerged from the articulation of a problem, which may have raised issues about whether time and cost go hand-in-hand; and therefore whether the majority of energy would be better expended controlling time rather than, as seems at present, controlling cost.

Raw data for time predictability percentages are not provided in Table 8.26 for brevity; the method of calculating percentages could be simply: actual duration − contract duration/ contract duration × 100. Thus, for a project with an actual duration of 55 weeks, and contract

**Table 8.25** Data for the correlation test.

| | 1 Method of procurement; measurement level categorical. 1 = competitive, 10 = collaborative | 2 Method of procurement; measurement level improved to interval. Anchored with 0 = competitive and 10 = collaborative | 3 Pre-start percentage change in cost predictability |
|---|---|---|---|
| 1 | 2 | 6 | +5.0 |
| 2 | 2 | 6 | 0 |
| 3 | 1 | 5 | 0 |
| 4 | 1 | 5 | −1.1 |
| 5 | 1 | 4 | +15.3 |
| 6 | 2 | 8 | +2.5 |
| 7 | 1 | 3 | +1.0 |
| 8 | 1 | 3 | +8.3 |
| 9 | 1 | 2 | +6.2 |
| 10 | 2 | 9 | +4.8 |
| 11 | 2 | 9 | −3.0 |
| 12 | 1 | 2 | +4.5 |
| 13 | 2 | 10 | −2.8 |
| 14 | 1 | 5 | −7.2 |
| 15 | 2 | 7 | −1.0 |
| 16 | 1 | 3 | 0 |
| 17 | 2 | 5 | −0.1 |
| 18 | 1 | 4 | 0 |
| 19 | 1 | 5 | +6.6 |
| 20 | 1 | 4 | −0.2 |
| 21 | 2 | 6 | −2.2 |
| 22 | 2 | 8 | −5.2 |
| 23 | 1 | 3 | +8.2 |
| 24 | 1 | 2 | +2.0 |
| 25 | 1 | 2 | +1.2 |
| 26 | 1 | 2 | +3.2 |
| 27 | 1 | 4 | +10.0 |
| 28 | 2 | 7 | −3.0 |
| 29 | 2 | 8 | −2.2 |
| 30 | 2 | 6 | −2.9 |
| Total | | | 47.90 |
| Mean | | | 1.60 |

duration of 50 weeks, the calculation would be: $55 - 50/50 \times 100 = 10\%$. The scatter diagram is the stage 1 analysis, and is illustrated in Figure 8.4.

Eyeball observation suggests a positive relationship; if the points on the graph were encircled by a rugby ball, that ball would be 'thinnish'. The relationship does appear to be linear; not curved or parabolic. There are no real outliers; the closest outlier of interest is the project that beat its time target, but ran over its cost budget. The table identifies this was project 19, which beat its time target by 1%, but ran over budget by 6%. It is not so strong an outlier that it should be removed from the analysis.Observation of the Spearman formula shows it is in five parts. Part 1 of $6 \, \Sigma D^2$, is presented in Table 8.26. The table allows the ranking scores to be determined easily (see example in Table 7.6). Note that in the ranking procedure,

**Table 8.26** Spearman correlation test.

| | 1<br>Pre-start percentage change in cost predictability | 2<br>Rank cost predictability | 3<br>Pre-start percentage change in time predictability | 4<br>Rank time predictability | 5<br>D<br>(col 3 − col 4) | 6<br>$D^2$<br>(col 5 squared) |
|---|---|---|---|---|---|---|
| 1 | +5.0 | 24 | +1.0 | 13 | 11 | 121 |
| 2 | 0 | 14.5 | +5.0 | 20 | −5.5 | 30.25 |
| 3 | 0 | 14.5 | 0 | 8 | 6.5 | 42.25 |
| 4 | −1.1 | 10 | +1.1 | 14 | −4 | 16 |
| 5 | +15.3 | 30 | +10.5 | 26 | 4 | 16 |
| 6 | +2.5 | 20 | 0 | 8 | 12 | 144 |
| 7 | +1.0 | 17 | 0 | 8 | 9 | 81 |
| 8 | +8.3 | 28 | +12.5 | 28 | 0 | 0 |
| 9 | +6.2 | 25 | +3.0 | 18 | 7 | 49 |
| 10 | +4.8 | 23 | +12.0 | 27 | −4 | 16 |
| 11 | −3.0 | 3.5 | 0 | 8 | −4.5 | 20.25 |
| 12 | +4.5 | 22 | +8.5 | 25 | −3 | 9 |
| 13 | −2.8 | 6 | 0 | 8 | −2 | 4 |
| 14 | −7.2 | 1 | −2.0 | 2 | −1 | 1 |
| 15 | −1.0 | 10 | 0 | 8 | 2 | 4 |
| 16 | 0 | 14.5 | −3.3 | 1 | 13.5 | 182.25 |
| 17 | −0.1 | 12 | +2.1 | 17 | −5 | 25 |
| 18 | 0 | 14.5 | +5.0 | 20 | −5.5 | 30.25 |
| 19 | +6.6 | 26 | −1.0 | 3 | 23 | 529 |
| 20 | −0.2 | 11 | 0 | 8 | 3 | 9 |
| 21 | −2.2 | 7.5 | +5.0 | 20 | −12.5 | 156.25 |
| 22 | −5.2 | 2 | 0 | 8 | −6 | 36 |
| 23 | +8.2 | 27 | +15.0 | 29 | −2 | 4 |
| 24 | +2.0 | 19 | +8.0 | 24 | −5 | 25 |
| 25 | +1.2 | 18 | +2.0 | 15 | 3 | 9 |
| 26 | +3.2 | 21 | +5.0 | 20 | 1 | 1 |
| 27 | +10.0 | 29 | +20.0 | 30 | −1 | 1 |

*(continued)*

**Table 8.26** (*Continued*)

| | 1<br>Pre-start percentage change in cost predictability | 2<br>Rank cost predictability | 3<br>Pre-start percentage change in time predictability | 4<br>Rank time predictability | 5<br>D (col 3 − col 4) | 6<br>$D^2$ (col 5 squared) |
|---|---|---|---|---|---|---|
| 28 | −3.0 | 3.5 | +2.0 | 15 | −11.5 | 132.25 |
| 29 | −2.2 | 7.5 | +3.0 | 18 | −10.5 | 110.25 |
| 30 | −2.9 | 5 | 0 | 8 | −3 | 9 |
| Total | 47.90 | | 114.4 | | | 1813 |
| Mean cols 1 and 3 | 1.60 | | 3.81 | | | $6\Sigma D^2 = 6 \times 1813 = 10878$ |
| SD | 4.88 | | 5.52 | | | |

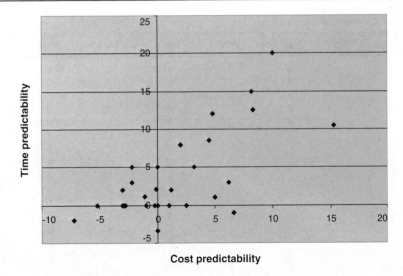

**Figure 8.4** Scatter diagram for time and cost predictability.

the two separate groups are treated separately (not as one group, as in the Mann–Whitney test). Note the ranking procedure for time predictability in column 4 shows that nine projects finish on time, and are thus equal rank position 4. These projects take up the positions 4 to 12; the mean rank for the Spearman calculation between 4 and 12 is 8, as shown:

- Part 1 is summarised in column 6. The sum of differences squared ($\Sigma D^2$) is 1813, and 6 $\Sigma D^2 = 10878$
- Part 2: $(N^3 - N) = (30 \times 30 \times 30) - 30 = 27\,000 - 30 = 26970$
- Part 3: $10878/26970 = 0.403 = $ say 0.40
- Part 4: $1.00 - 0.40 = 0.60$
- Part 5: determining whether the correlation coefficient of 0.60 arose due to chance. From the table in appendix L, the critical value of Spearman's rho with the significance level set at $0.05 = 0.364$. The calculated value of 0.60 must be equal to or more than the critical value. The value 0.60 is more than 0.364, therefore the finding is: reject the null hypothesis with $p$ set at $\leq 0.05$; there is a relationship between time predictability and cost predictability. The correlation coefficient is 0.60; this can be classified as a moderate to good relationship. Projects that beat time targets may beat time budgets. Projects that complete on time may complete to budget. Projects that overrun in terms of time may also overrun budget.

$R^2$ is 0.36; therefore, 36% of the variation in time is related to the variation in cost. The parametric Pearson correlation coefficient calculated by Excel $= 0.692$; that contrasts slightly with the non-parametric Spearman figure.

## The consequence of larger sample size and a wider spread of data

As with the chi-square and difference in means tests, sample size affects $p$ value. Take for example hypothetical data in Table 8.27; there is no need to attach meaning to the

**Table 8.27** Hypothetical data.

| | Var1 | Var2 | Var3 | Var4 | Var5 | Var6 |
|---|---|---|---|---|---|---|
| 1 | 8 | 6 | 8 | 6 | 9 | 2 |
| 2 | 4 | 7 | 4 | 7 | 5 | 4 |
| 3 | 6 | 9 | 6 | 9 | 7 | 9 |
| 4 | 2 | 10 | 2 | 10 | 6 | 10 |
| 5 | 4 | 9 | 4 | 9 | 5 | 11 |
| 6 | | | 7 | 6 | 8 | 4 |
| 7 | | | 6 | 7 | 7 | 4 |
| 8 | | | 6 | 9 | 6 | 11 |
| 9 | | | 2 | 10 | 2 | 7 |
| 10 | | | 4 | 9 | 4 | 11 |
| 11 | | | 8 | 7 | 10 | 5 |
| 12 | | | 4 | 7 | 0 | 6 |
| 13 | | | 7 | 9 | 0 | 8 |
| 14 | | | 2 | 10 | 2 | 11 |
| 15 | | | 2 | 9 | 6 | 8 |
| Count or 'n' | 5 | 5 | 15 | 15 | 15 | 15 |
| Range | 6 | 4 | 6 | 4 | 6 | 6 |
| Maximum | 8 | 10 | 15 | 15 | 15 | 15 |
| Minimum | 2 | 6 | 2 | 6 | 0 | 2 |
| Mean | 4.80 | 8.20 | 4.80 | 8.20 | 5.13 | 7.40 |
| Median | 4 | 9 | 4 | 9 | 6 | 8 |
| Mode | 4 | 9 | 4 | 9 | 6 | 11 |
| Standard deviation | 2.28 | 1.64 | 2.21 | 1.44 | 3.04 | 3.09 |
| Correlation coefficient | $-0.72$ | | $-0.68$ | | $-0.38$ | |
| P value | $>0.05$ | | $\leq 0.05$ | | $\leq 0.05$ | |
| Finding | Cannot reject | | Reject | | Reject | |

variables. The scatter diagram for Var1 and Var2, in Figure 8.5a, shows potential for a relationship between the variables. In this case it is shown as negative. However, the sample size is so small, it could obviously have arisen due to chance; even a robust sampling methodology, with such small numbers, could not reasonably be expected to select a sample that reflects the population. The sample may have, by fluke, picked data that show high scores on Var1 with low scores on Var2; also medium scores on Var1 with medium on Var2; finally, low scores on Var1 with high scores on Var2. There may reasonably be elsewhere in the population data that show a different perspective, e.g. low scores on Var1 with low scores on Var2 and high scores on Var1 with high scores on Var2. This is confirmed by a correlation coefficient of -0.72 and, very importantly, a $p$ value calculated as $>0.05$.

Now take the scatter diagram for Var3 and Var4 in Figure 8.5b; the measures of central tendency and spread are similar to Vars 1 and 2. Now the sample size is so large that the apparent relationship should not reasonably have arisen due to chance. This is confirmed by a $p$ value calculated as $\leq 0.05$ and a similar correlation coefficient of $-0.68$

The scatter diagram for columns Vars5 and 6 (Figure 8.5c) has a wider spread, as noted by the standard deviations. The consequence is to create a wider rugby ball; the correlation coefficient reduces to $-0.38$.

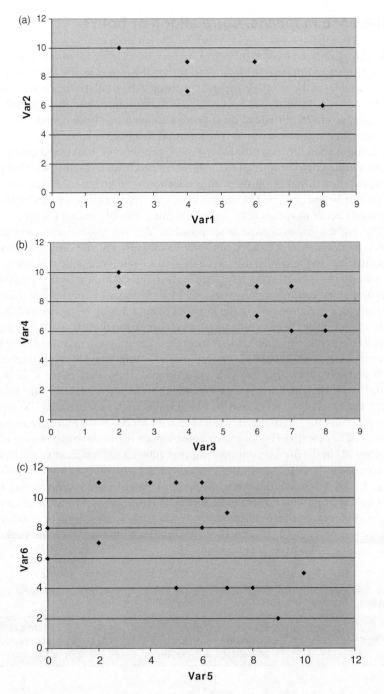

**Figure 8.5** Scatter diagrams of (a) var1 and var2, $n = 5$, CC $= -0.72$, and $p$ value $>0.05$, (b) var3 and var4, $n = 15$, CC $= -0.68$, and $p$ value $\leq 0.05$ and (c) var5 and var6, $n = 15$, CC $= -0.38$, and $p$ value $\leq 0.05$.

## 8.9 Difference in means, correlations or both?

When designing data collection processes, it may be helpful to consider similarities in correlation studies to unrelated difference in means tests, such as the Mann–Whitney test. One of the variables in the Mann–Whitney is in groups; the similarity arises if the two groups required for a Mann–Whitney are opened up on a continuum, into three or better four, or better five, six, seven, etc. Provided these groups are on a continuum, and not categorical/nominal data, they are ordinal or interval data, and therefore suitable for a correlation test. In research about smoking, two groups, group 'A' people who smoke, and group 'B' people who do not smoke, could be opened to the interval level by determining whether people are zero, light, medium or heavy smokers. In gender research, group 'A' males, and group 'B' females, cannot be opened up. The important issue here is, that since correlation tests are richer than difference in means tests, when designing the data collection process, measure data at the highest measurement level possible, e.g. if smoking can be measured at the interval level, do so; but gender can only be measured at the categorical level. There are also similarities to the related difference in means tests, if the same variable is measured twice, but in different contexts, e.g. measuring variables within a time frame, but in two different companies, countries, industries, etc. Indeed, it may be appropriate to undertake a related *t* test and correlation test on the same set of data. It is to be recognised that the two tests do different things; the basis for a test should come from the articulation of the problem and an objective/hypothesis should be set from this. The problem may justify one test or the other. However, researchers stumble on important findings accidentally. Since the data are there, it is perfectly reasonable that you 'poke around' those data to see if there are other interesting issues that may arise. Take for example a set of data briefly presented in Table 8.28. The basis of the study may be to examine differences between UK construction unemployment and the rest of the EU as an argument for closer (or more distant) alignment of economies. The objective is to 'compare construction unemployment rates in the UK and the rest of the EU in the decade commencing year 2000'. As a comparative study, a related *t* test will identify if there are significant differences. As part of 'post hoc' analysis, or 'data snooping', it may be appropriate to see if unemployment rates in one impact on unemployment rates in the other. Is there a negative correlation, whereby high unemployment in one is a precept for migration to the other? Is there a basis for looking at data to compare the UK with just one other country in the EU? The difference in means tests and the correlation tests meet different objectives with the same data.

**Table 8.28** Example to illustrate data that may be use in comparative difference in means test, and in correlation test.

| Months and year | EU construction unemployment as a percentage of workforce | UK construction unemployment as a percentage of workforce |
| --- | --- | --- |
| June 2009 | 9.4 | 7.8 |
| May 2009 | 9.3 | 7.7 |
| April 2009 | 9.2 | 7.5 |
| March 2009 | 9.1 | 7.2 |
| Etc. to June 2000 | | |

Alternatively consider the related data in Table 8.22 about cost predictability. The two variables are percentage change on pre-tender budget, and percentage change on pre-start budget. A difference in means test is appropriate; a correlation test is inappropriate. There is no possibility that percentage accuracy at pre-tender can impact on percentage accuracy at pre-start. Whilst a correlation test will manipulate the numbers in the two columns and give a correlation coefficient, whatever the result, it would be meaningless.

## 8.10 Using correlation coefficients to measure internal reliability and validity

Baskets of questions or multiple item scales inherently improve reliability. Single item questions about one issue are unreliable. It is better to use a basket of questions that combine to measure one variable. Consider a study founded in a problem around quality of workmanship of craftspeople. It is decided to use a basket of questions to measure the variable 'commitment to quality'. In this example, a small basket of five questions is illustrated, self designed, each with five possible answers. Since it is self designed and not a questionnaire validated by previous studies, it is important to measure its reliability. Questions 2 and 5 have their poles reversed to minimise the effect of yea-sayers and nay-sayers (Chapter 5). The numerical code for each response is indicated against each question, though you would not necessarily show these codes to respondents on the questionnaire. The questions are:

(1)  How inclined are you to rework your product if it is faulty? Intensely inclined (4)/very inclined (3)/inclined (2)/not very inclined (1)/not inclined (0)
(2)  How carefully do you check your own work when it is completed? Do not check (0)/not very closely (1)/closely (2)/very closely (3)/extremely closely (4)
(3)  How often do you criticise your colleagues' work? Always (4)/sometimes (3)/not often (2)/rarely (1)/never (0)
(4)  How attentive to detail are you in your work? Intensely attentive (4)/very attentive (3)/attentive (2)/not very attentive (1)/not attentive (0)
(5)  How often do you consult your supervisor if you see quality problems? Never (0)/rarely (1)/not often (2)/sometimes (3)/always (4)

A small data set, with $n = 10$, is shown in Table 8.29. Adjustment for yea-sayers and nay-sayers is made in questions 2 and 5 by changing response codes thus: 0 becomes 4, 3 becomes 1, 2 stays as 2, 1 becomes 3, and 0 becomes 4. Scatter diagrams can be plotted for each individual question against the total summed score for remaining questions, e.g. Q1 plotted against the total summed score of questions 2, 3, 4 and 5. If each individual question in the basket is truly a measure of this variable, the following pattern will be evident in the scatter diagram:

• Respondents who score highly for one question will score highly for others
• Respondents who score mid-range for one question will score mid-range for others
• Respondents who score low for one question will score low for others

Whether or not this occurs can be illustrated on a scatter diagram, and then by calculating correlation coefficients.

**Table 8.29** Data set for a measure of 'commitment to quality', $n = 10$.

| | Q1 | Q2 | Q3 | Q4 | Q5 | Q2R | Q5R | TOT | Total % | Total ex Q1 | Total ex Q2R | Total ex Q3 | Total ex Q4 | Total ex Q5R |
|---|---|---|---|---|---|---|---|---|---|---|---|---|---|---|
| 1 | 1 | 2 | 2 | 3 | 1 | 2 | 3 | 11 | 55 | 10 | 9 | 8 | 9 | 8 |
| 2 | 2 | 1 | 0 | 3 | 1 | 3 | 3 | 11 | 55 | 9 | 11 | 8 | 8 | 8 |
| 3 | 1 | 3 | 2 | 1 | 3 | 1 | 1 | 6 | 30 | 5 | 4 | 5 | 5 | 5 |
| 4 | 1 | 3 | 4 | 2 | 3 | 1 | 1 | 9 | 45 | 8 | 5 | 7 | 8 | 8 |
| 5 | 4 | 1 | 4 | 3 | 0 | 3 | 4 | 18 | 90 | 14 | 14 | 15 | 15 | 14 |
| 6 | 4 | 0 | 2 | 4 | 0 | 4 | 4 | 18 | 90 | 14 | 16 | 14 | 14 | 14 |
| 7 | 0 | 1 | 2 | 3 | 1 | 3 | 3 | 11 | 55 | 11 | 9 | 8 | 8 | 8 |
| 8 | 1 | 2 | 1 | 2 | 2 | 2 | 2 | 8 | 40 | 7 | 7 | 6 | 6 | 6 |
| 9 | 0 | 0 | 4 | 1 | 3 | 0 | 1 | 6 | 30 | 6 | 2 | 5 | 6 | 5 |
| 10 | 1 | 3 | 3 | 1 | 1 | 3 | 3 | 9 | 45 | 8 | 6 | 8 | 8 | 8 |
| | | | | | | | | | | 0.75 | 0.95 | 0.15 | 0.73 | 0.77 |

Correlation coefficients, e.g. 0.75 is the correlation coefficient between Q1 and total ex Q1, 0.95 is the correlation coefficient between Q2R and total ex Q2R,

Key: ex = excluding. R = reversed. For example the column labelled Total 1 ex Q1 is the summed total of Q2R + Q3 + Q4 + Q5R. For respondent 1 this is $2 + 2 + 3 + 3 = 10$

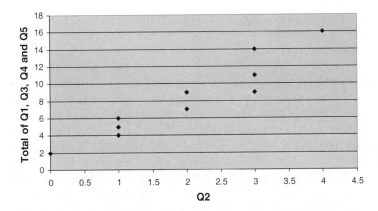

**Figure 8.6** Internal reliability check: Q2 versus total of Q1, Q3, Q4 and Q5.

Figures 8.6 and 8.7 plot the scores obtained to Q2R on the horizontal scale versus the total for all other questions (1, 3, 4 and 5) on the vertical scale, and similarly Q3. Eyeball observation of the scatter diagrams suggests:

- From Figure 8.6 it can be observed Q2 is a reliable question; it appears that respondents who score high (or mid-range or low) for commitment to quality on this question, do similarly for other questions summated together. This is confirmed by the correlation coefficient calculated by Excel to be 0.95; from the table in Appendix L this is significant with $p$ set at 0.05
- From Figure 8.7 it can be observed Q3 is not a reliable question; it appears that respondents who score high (or mid-range or low) for commitment to quality on this question, do not score similarly for other questions summated together. This is confirmed by the correlation coefficient calculated by Excel to be 0.15; from the table in Appendix L this is not significant with $p$ set at 0.05.

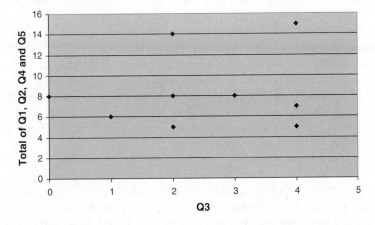

**Figure 8.7** Internal reliability check: Q3 versus total of Q1, Q2, Q4 and Q5.

Scatter plots should be drawn for all other individual questions. However, if placed in the main body of the document, this could become monotonous. Therefore, perhaps report the correlation coefficients in the main body and place scatter diagrams in an appendix.

You must try and think through why Q3 may be an unreliable question; it does not appear to be measuring the same concept as the other four questions. Is it the case that people, though perhaps committed to quality, do not put this ahead of criticising colleagues, with all the potential consequences that that may have? If you have piloted the questionnaire, you may change this question before the main study is administered. If you have not piloted it, you should write about how this unreliable question may have skewed your results, or alternatively you may prefer to conduct the main analysis, with answers to this question excluded.

If you have used a gold standard questionnaire, on the one hand it may be argued that such tests are not necessary since they have already been done. However, it is appropriate that you still perform them. Perhaps the questionnaire has not been tested in the context of a study similar to yours; also by doing the tests it is an opportunity for you to demonstrate your analytical skills and understanding of reliability.

Using baskets of questions or multiple item scales inherently improves reliability, or more particularly the repeatability aspect of reliability in questionnaires. Suppose you used only one question to measure the variable 'commitment to quality'. This question could be 'how committed are you to producing high quality work?' You may provide five possible answers, coded numerically 0, 1, 2, 3 and 4: intensely committed (4)/very committed (3)/committed (2)/not very committed (1)/not committed (0). On a 100 point wide percentage scale these five numbers are replicated by 0, 25, 50, 75 and 100 (e.g. not very committed = 1 out of 4 = 25%). Imagine that you wanted to test the reliability of your questionnaire through repeatability, though in practice you are unlikely to want to do this. At the first point of answering the respondent is in a good mood, and indicates the qualitative answer 'very committed', that you code numerically as 3 or 75%. One month later, you administer the same question to the same respondent. The respondent is still in a good mood, but by chance indicates the qualitative answer 'committed' that you code numerically as 2 or 50%. A 25% difference, or if we manipulate the figures, 75 is actually 50% higher than 50 and, therefore, potentially 50% different. This is clearly not acceptable; the measure of the variable is not repeated.

However, using a basket of questions, at the first point of answering the respondent scores 1, 2, 3, 3 and 4: a total of 13 out of a maximum of 20, or 65%. One month later, by chance the respondent gives a different response to the first three of the five questions, thus 2, 3, 2, 3 and 4. The total is 14 out of 20, or 70%. Changes in two of the questions have been in the opposite direction and have cancelled each other out. There is now only a 5% difference. The larger the basket of questions, the more likely you will minimise potential for large differences, and thus your questionnaire will be more repeatable.

Correlation coefficients can also be used in studies that specifically set out to measure validity between variables. Take the potential relationship between the variables 'performance at interview' (IV) and 'future likely performance in the workplace' (DV). This is in the context of interviews being used in the selection process for jobs. The question is 'are interviews a valid way of predicting future likely performance in the workplace?' If the IV and DV correlate well together, the answer is yes. Research has been

**Table 8.30** Summary of inferential statistical results.

| Test no. | IV or DV | Variable name | Test type | Alpha level (sig) | Results | p value | Reject or cannot reject the null |
|---|---|---|---|---|---|---|---|
| 1 | IV | Method of procurement; competitive or collaborate | Chi-square | 0.05 | See contingency table | >0.05 | Cannot reject |
|  | DV | Cost predictability |  |  |  |  |  |
| 2 | IV | Method of procurement; competitive or collaborate | Mann–Whitney | 0.05 | Mean scores; competitive = +3.41%, collaborative = −0.78%. Overall mean = +1.6% | ≤0.05 | Reject |
|  | DV | Cost predictability |  |  |  |  |  |
| 3 | IV | Stage of estimate; pretender or prestart | Wilcoxon | 0.05 | Mean scores; pretender % change = +7.36%, prestart % change = −1.60% | ≤0.01 | Reject |
|  | DV | Cost predictability |  |  |  |  |  |
| 4 | IV | Stage of estimate; pretender or prestart | Related t test | 0.05 | Mean scores; pretender % change = +7.36%, prestart % change = −1.60% | ≤0.01 | Reject |
|  | DV | Cost predictability |  |  |  |  |  |
| 5 | IV | Method of procurement; competitive or collaborate | Pearson's correlation | 0.05 | Mean scores; method of procurement = 5.1, cost predictability = +1.6%. Correlation coefficient = −0.47 | ≤0.05 | Reject |
|  | DV | Cost predictability |  |  |  |  |  |
| 6 | IV | Cost predictability | Spearman's correlation | 0.05 | Mean scores; cost predictability = +1.6%, time predictability = +3.81%. Correlation coefficient = 0.60 | ≤0.05 | Reject |
|  | DV | Time predictability |  |  |  |  |  |

undertaken on this issue in other disciplines, and it has been found not to be the case. The correlation coefficient was only 0.40; a figure much closer to 1.00 would be expected if the interview were a valid measure of future likely job performance. It is possible to imagine people who perform poorly at interview, perhaps because they are nervous, but perform really well at work. Similarly, there will be people who make a good impression at interview, but do not perform. Because the simple interview is known to be a poor predictor of job performance, companies seek to introduce other measures into the selection process, e.g. references (although these are notoriously invalid measures, especially when interviewees name referees), panel and two-stage interviews, intelligence tests, and personality tests.

## 8.11 Summarising results

If your analysis section includes a lot of tests, it may be helpful to readers if they are summarised in one table at the end the dissertation chapter. This book chapter includes many tests; they are summarised in Table 8.30.

## Summary of this chapter

Inferential statistics are more complex than descriptive statistics, but if you are able to execute such tests, this gives you the opportunity to gain more marks. Do not attempt these tests if you do not have a firm understanding of the underlying principles. The inferential tests can be put on the continuum of: chi-square, difference in means and correlation. The appropriate test is driven by the level of the data for the IV and the DV, and this is similarly on a continuum of categorical/nominal, ordinal and interval/ratio. You should identify whether your data set is non-parametric or parametric. Correlation tests can be used to test internal reliability and validity. Since you may perform many statistical tests on your data, it is useful to summarise them in a table at the end of your chapter.

# 9 Discussion, conclusions, recommendations and appendices

The objectives of each section of this chapter are:

9.1 Introduction: to provide context for the final part of the study
9.2 Discussion: to explain the expected contents
9.3 Conclusions: to explain the expected contents
9.4 Recommendations: to explain the expected contents
9.5 Appendices: to identify the type of material to be included
9.6 The examiner's perspective: to give an overview of the dissertation

## 9.1 Introduction

The final two chapters may be a discussion and, integrated into one chapter, the conclusions and recommendations. Together these two chapters may account for one third of your dissertation. You should make sure that the material included meets the expectations of an undergraduate dissertation, and ensure that you keep these chapters focused on your objectives.

A discussion should bring together the literature, findings and results from all the data that you have collected and analysed, and develop the critical appraisal that you have undertaken in the literature review. Conclusions may state what needs to happen or explain cause and effect issues. Your recommendations should come from your data, and should not be too ambitious nor suggest radical change in government or industry direction that is unlikely to materialise. To recommend another study may be appropriate. Appendices should include all the material that is not necessary to meet study objectives. At the end of your study you should undertake your own appraisal, and recognise the limitations of what you have done, including judgements about reliability and validity. You will need to do a lot of early reading and make sure you write as you go.

*Writing a Built Environment Dissertation: Practical Guidance and Examples.* Peter Farrell.
©2011 Peter Farrell. Published 2011 by Blackwell Publishing Ltd.

## 9.2 Discussion

The discussion should be focused on aims and objectives. Perhaps it should be articulated under the heading of each. Discussion chapters in dissertations are sometimes weak or even non-existent; they go straight from results to conclusions. It is as though students run out of time, and they are exhausted after completing the analysis. Others label the final chapter 'discussion, conclusions and recommendations', but it is only a title, and little real discussion takes place. Dissertations that appear to be destined for good marks in the early and middle parts, weaken towards the end. You should view the discussion as an opportunity to make the dissertation destined for good marks into one that deserves excellent marks. Though time management is an important issue (you need to complete the analysis several months before the final submission date), there is also a need to focus upon what good discussions should comprise. *The Oxford Dictionary* definition of a discussion is:

- An examination by argument
- A debate

The discussion requires that results and findings be evaluated against each other and against the theory and the literature. You should integrate what you have found in your data with what the literature states. No more reading is necessary; just go back to your literature chapter and bring forward the same sources. Cite them again, e.g. 'this study found that photovoltaic cells have a payback period of 30 years. This is supported by the work of Smith (2010) and Brown (2008), but it is contrary to the work of Baker (2006), who though using a different methodology found a shorter period of 20 years'.

It is in this chapter that you can take some imaginative leaps in developing your findings. To the extent that this study is 'you', this is where you can demonstrate your insight. But you should also get some help, to avoid the danger of viewing your findings too narrowly. There is huge potential to do some qualitative work towards the latter part of your study that could gain extra marks. The thrust of some informal interviews could be to report what you have found and ask respondents what they think.

The findings are based on what the data tells you; they are 'fact'. Now that you have (or have not) found relationships between variables, the discussion gives you the opportunity to carefully speculate about important issues such as 'why' and 'how', which are important parts of theory building. If the discussion is poor, findings about relationships are effectively left isolated. There should be links into any theories identified in earlier parts of the dissertation. To illustrate that your speculation is careful, use some caveats such as 'perhaps' or 'possibly'. You should demonstrate intelligent interpretation, with insight. Then, start to develop new hypotheses as recommendations for further study. In laboratory work, if you have found that material 'A' performs better than 'B', what are the scientific reasons for this—does the answer lie in understanding chemical behaviour of materials—why? If you find that construction professionals are less stressed working on projects procured using partnering methods, why is that? How can other methods of procurement learn from partnering to reduce stress levels?

Discussion should deduce explanations for similarities and differences and examine potential causes or potential variables. The discussion should start to open the door for what will come in the conclusion, perhaps in the form of stating what needs to happen.

Conclusions should not just 'pop out' in the conclusion chapter. They should be developed in the discussion chapter, and any arguments supporting them should be articulated. You should have the aspiration to do something original in your work, in a very modest way. The discussion should tease out what you have done that has been different, and how it may have implications for future studies. Provided your work is well founded and structured, however modest it is, it has its place in the knowledge base of your chosen subject area. Do not be afraid to emphasise this.

You should not have come into your research with intransigent preconceived ideas about what outcomes will be. You may have had perceptions and even 'hopes', but you must let the data give you answers. Some students use the discussion to embark on their own personal treatise about how to solve the problems of industry. Do not do this; there are analogies similar to the literature review. The discussion is a weighing exercise where you bring the evidence to bear, and you sensibly recognise the weight of your experience in the context of the experience and research of others detailed in literature. You must give your judgement, but make sure it is based on all the evidence, and not unreasonably biased by your own experience. The evidence before you is the literature and the new data collected and analysed in your study. Only stray from the evidence with careful speculation.

There are many issues in life where there are two sides to the debate. You must let the argument take place. The political world is full of issues where the decision making process is often controversial, with seemingly compelling evidence on each side of an argument. Judgements are often complex, and it may be the case that you have to articulate that in one set of market conditions the answer is 'abc', but perhaps in another 'xyz'. Many issues in life are not black or white; they are grey. A discussion should be written with sensitivity, recognising blurred boundaries that exist. It must try to have empathy with and respect for what others may judge. Whilst you may make a judgement on one side, the discussion chapter should not leave readers to examine the arguments from the other side. In your writing you must recognise both sides, even if you are overwhelmingly convinced by the evidence in one direction. When making your judgements, it may be that others, including your examiner, disagree. This is fine, and you will not be marked down, providing you are able to defend your judgements based on correct interpretation of the evidence.

You should also consider the cynic who may be critical of the research. It is useful to articulate what the cynic may find, recognising the limitations of what you have done. If your work is qualitative, you may choose not to have a separate discussion chapter. The discussion may be embedded in the narrative that ends your analysis, results and findings chapter. If this is your choice, fine, but do make sure that the narrative is of the appropriate weight, and includes all the features of a discussion as though it were a separate chapter. Alternatively, you may wish to take the narrative out of the analysis, results and findings, and place it as a separate chapter with the discussion.

## 9.3 Conclusions and recommendations

The conclusion brings together the whole of the dissertation. A conclusion should be a logical outcome of all that has gone before. Conclusions are often weak, sometimes comprising merely one page of writing not related to objectives. Students run out of time, and are exhausted after completing the analysis. Students may write conclusions the day before submission is due; this must not be the case. The conclusions must be laboured over long and hard. They should be the result of many iterations, approached with 'insight' and given a great deal of thought. Dissertations that appear to be destined for good marks in the early and middle parts, weaken or even worse, fail at this point. No matter how good the early work in a dissertation, it is impossible to pass without substance in conclusions—absolutely impossible. Time management is again important, but there is a need to focus upon what good conclusions should comprise. *The Oxford Dictionary* definition of a conclusion is:

- A final result; a termination
- A judgement reached by reasoning
- The summing up of an argument, article or book
- A proposition that is reached from given premises

The purpose of academic study is to come up with some really good conclusions. To reiterate, the whole study must hang around the objectives. In the introduction chapter you have stated your research questions and have moved through your document in a structured sort of way to arrive at some answers. The objectives imitate questions. In the same way that previous chapters have been focused on aims and objectives, so too should the conclusion. The conclusion closes the study out, but links it back to the aim and objectives in the introduction. The strap line for objectives could be given as separate headings to make sure you do not lose sight of them. Number conclusions to imitate numbering of objectives.

There should be no new material and no quotations from the literature. As a chapter, the conclusion should stand alone, and readers should be able to understand concepts behind each conclusion without having to refer to earlier chapters. Therefore, at the risk of being monotonous, key parts of earlier chapters will be repeated. Be mindful that some readers will not take the full journey through your dissertation. To help them, restate important parts of the literature and summarise the method, including populations and samples. Repeat your results and findings. This may include some key numbers such as mean scores or correlation coefficients. Important elements of the discussion should also be summarised succinctly. Judgements, which have started their formation in earlier chapters, can now be asserted firmly. Some words may be very similar to those used in the abstract.

What is a conclusion? Studies executed on the premise of establishing causes may be appropriate to a conclusion writing style which suggests something needs to change or happen; that is to determine what the IVs are and to conclude that they need to be manipulated. However, some studies may not be appropriate for this style. Try to answer cynically-put questions; 'now that you have these results and findings, so what? Where do we go from here?' If it is found that 'A' is better than 'B', you may draft conclusions on the assumption that something needs to change or something needs to happen. Either 'B' needs to improve, or 'A' should be used exclusively at the expense of 'B'. The objective of some research may be to develop theories or explanations as to why things happen, and look at the cause and effect issues. The explanation may not be clear cut and definitive, and therefore a conclusion may be substantive in its own right if it merely articulates why something happens. Such explanation needs to be academically robust, and stem from the data analysis, results and findings in the study. In the conclusion also extend the 'how' theories (or better labelled hypotheses) that are started in the discussion. If the evidence to support findings is overwhelming, the tone of conclusions can be assertive. If more work is needed, the tone of conclusions may be more tentative.

An opportunity to gain marks is available through a carefully written section under a sub-heading of 'limitations and criticisms of the study'. You should not let readers spot weaknesses in your study. The undergraduate dissertation process is only training for research, and to give you an appreciation of the role of research in society. This appreciation is enhanced if you know where you went wrong. If you went wrong, this is understandable; but if you went wrong and do not know it, then you should not score high marks. Some weaknesses in your study may be where you made mistakes or misunderstood something. Some weaknesses may be due to failings in time management, which you reflect should not have occurred. Do not be afraid to mention these. You may, however, have completed a dissertation within the reasonable bounds of what universities can expect. Some fundamental weaknesses in the study may have been unavoidable, due to lack of resources. It is here where you should reflect on how valid your work is. Suppose it were appropriate to measure validity on a scale of 0–10, with 10 the highest validity, and in your undergraduate dissertation it was only reasonable to expect you to score 5/10. If you do indeed score 5/10, you should score excellent marks; but if you score 5/10 and you can articulate why you could not score 10/10 and how 10/10 could be scored, you will get even more marks.

Recommendations follow conclusions. Some students do not write recommendations, whilst others use them to put right all the ills of industry, unrelated to study objectives. You should not write recommendations that are 'hunches' or personal 'hobby horses'; they must come from data in studies. Do not be too ambitious in writing recommendations, and do not write too many. Recommendations in authoritative reports written for governments some-times fail at implementation. Your recommendations will similarly fail if action would cost substantial resources. Recommendations should reflect objectives and therefore implicitly reflect conclusions.

If you have compelling findings that would benefit industry, you can write your recom-mendations for that audience. However, it is more likely to be the case that the appropriate audience is academia. On this basis, it may be better to write your recommendations as though suggesting more study. This fits well with the scenario of theory building, testing and re-testing. Recommendations for more study should be well thought out and fully developed. Therefore, for one or two pages, describe the problem, and state the research questions, objectives and hypotheses. Define the variables and state how they should be measured. Describe populations, sampling method and analytical method. This is potentially a lot of work, so it is probably sufficient that you do it for just one objective. It may be that as part of your literature review you noted an important, related topic area, but you decided not to formulate an objective for it since you were limited by resources. A recommendation for further study could be founded in such an area. There is the possibility that subsequent students could pick up your recommendations and carry it as a baton for their work. Having written the limitations and criticisms of your study, you may reflect with the benefit of hindsight that your study did not take you where you really wanted to go. This feeling is understandable, but do not get upset about it. The recommendations are a really good opportunity for you to write the full research proposal for the study you wish you had done.

## 9.4 Appendices

Appendix singular, appendices plural, provide supplementary information and necessary supporting evidence. They are not included in word counts. The following strap line may help, though not to be taken literally:

> 'if it should be in, it should be in; if it should be out, it should be out. If in doubt, put it in the appendices'

The context of 'if it should in' is, is it relevant to the objectives? All material that is necessary to meet study objectives should be in the main body of dissertations. Appendices should not be too long. If your study objective comprises a detailed appraisal of a British Standard, or a Standard Form of Building Contract, or a Government report, do not include the full document. If readers really want to access documents, they can do so by following leads in your citations and referencing. If there are one or two pages that are crucial to the objectives, by all means provide them for the convenience of readers. Appendices should be in the same order as they appear in the narrative, and they should be designated by a capital letter ('Appendix A', 'Appendix B', etc.) rather than by an Arabic number (1, 2, 3, etc.). It is useful to provide a header to appendices and label all pages A, B etc., to allow readers to locate them easily. New appendices should start on separate pages with page numbers following on from the references/bibliography sections. There should be a list of appendices in the preliminary pages at the front of the dissertation.

Appendices often include such things as survey instruments, questionnaires, verbatim transcripts of interviews, or photographs. Photographs may be of architectural features, defects in completed buildings, or failure modes of materials tested to destruction in laboratories. Those that are important to the objectives should be in the main body. It is often sufficient to include only summaries of raw data in the main body; the raw data sets themselves can usually be in an appendix. Results and findings should be in the main body. The analytical process may be assisted by computer. If you do your detailed analysis without software support, this may involve many calculations. A few calculations may be placed in the main body, but if you have reams of them, place them in the appendices. If you have used a questionnaire, do not include all those that are returned by respondents. Only a blank copy is required. If you have asked closed questions and converted qualitative responses to numbers (e.g. excellent = 4, very good = 3, etc.), also include another copy, with the numerical coding marked on it. A table that summarises the responses from any questionnaires should be presented in the main body.

During the proofreading process, you may reflect that some parts of the dissertation are too long, and you have overstepped the word guide. Overstepping may occur in the literature review, or you may have included too much detail in your analytical chapter. The material may not be directly relevant to your objectives. Rather than delete it and lose the material completely, include it in an appendix. Examiners may give you some credit for it. Examiners will not necessarily read your appendices, but they may browse them, perhaps if they have some doubts that need to be explored. Subsequent researchers may want to use details in appendices to develop their work, or to substantiate the validity of your work in their own minds.

## 9.5 The examiner's perspective

Examiner's are required to benchmark dissertations against carefully written and designed criteria. The criteria help to ensure that the mark that you are allocated is as objective, reliable and valid as possible. A test of marking validity would be that two markers who read a dissertation, independent of each other, allocate the same or nearly the same mark. There is some subjectivity involved in making most assessments (inevitably); the criteria are in place to move judgements along the continuum from subjectivity to objectivity, as far as is possible. The following examiner's perspective was adapted by Rudd (2005) from work by Brown and Adkins. It could be summarised around two generic themes; originality and validity. You may find it useful to guide your work from the start, in making your own judgements about the validity of your study, and finally in writing a section in your conclusion chapter about limitations and criticisms.

### Initial overview

- Read title and abstract for thesis, key idea, question, topic
- Turn to last chapter/conclusion to see how far this has been addressed
- Note any methodological weaknesses
- Turn to contents page to see if sufficient, appropriate evidence has been covered
- Read introduction for close definition of problem in context
- General impression? Formulate questions, based on introduction, as to what you would hope to be answered/discussed, e.g. congruency of methods with problem, possible sources of bias

### Review of literature

- To what extent is it relevant to the research issue?
- To what extent is it 'this is all and everything I know about this'?
- Is the review just descriptive, or analytical, evaluative?
- How well is technical/theoretical literature mastered?
- Are links between literature review and design of this study explicit?
- Is there a summary of the essential features of other work as it relates to this study?

### Design of study

- Precautions against likely sources of bias?
- Limitations of the design? Is candidate aware of them?
- Is the methodology for data collection appropriate?
- Are the techniques of analysis appropriate?
- Has the best design been chosen?
- Has the design been justified?

### Presentation of results

- Does the design appear to have worked satisfactorily?
- Have the hypotheses been tested?
- Do the solutions relate to the questions posed?
- Is the level and form of analysis appropriate to the data?
- Could the results have been presented more clearly?
- Are patterns and trends in results identified and summarised?
- Is a picture built up?

### Discussion and conclusions

- Is the candidate aware of possible limits in reliability/validity of the study?
- Have the main points in results been picked up and discussed?
- Are there links made with other literature?
- Is there an attempt to reconceptualise problems, to rethink/develop theory?
- Is there intelligent/imaginative speculation? Does it grow out of the results?

### Summative overview

- Is the standard of presentation adequate (English, style, notes, bibliography, etc.)?
- Is the thesis generally the candidate's own work?
- Does the candidate have general understanding of the field and how the thesis relates to it?
- Has the candidate thought through implications of findings?
- Is there evidence of originality?
- Does the study add to existing knowledge of this area?
- Is there evidence of development in research skills?
- Is it worth publishing in some form?
- Are any qualms/reservations found in the material, or do they arise from ideological/ epistemological differences?

## Summary of the dissertation process

Good time management and focus upon objectives is essential. The process that you follow needs to be underpinned by reading about your specialist subject area and about research methodology. Whilst you inevitably learn as you read, you need to have a reasonable grasp of both at an early stage to be able to set out a plan; therefore early reading is paramount.

A possible scenario for a study is set out:

- Write as you go; keep communicating with your supervisor
- Construct a well-founded description of the problem from which a provisional research question emerges
- The research question is translated into an objective

- Agree a protocol for working with your supervisor
- Complete your permission to do research form and information sheet in accordance with your university ethical code
- Undertake exploratory work (networking, informal or formal interviews) to redefine the problem, and reshape the question/objective
- Undertake the literature review
- Undertake more exploratory work to reshape the question/objective
- Establish the objective firmly; identify the variables
- Write the objective as a hypothesis
- Open-up the objective/hypothesis and its variables by defining them in the narrative in the context of your study
- Open-up the definition of the variables by designing a research instrument to measure them; the research instrument and its surrounding methodology must be the most appropriate instrument to meet the objective/measure the variables
- Collect the data and do the analysis
- Undertake further networking and informal or formal interviews to help shape the discussion and conclusions
- Employ extensive thought, interpretation and insight in writing discussion and conclusions
- Complete final write up and submit draft submission for feedback in good time
- Undertake final proofreading
- Submit ☺

## Summary of this chapter

A discussion should bring together the literature, findings and results from all data that you have collected and analysed, and develop the critical appraisal that you have undertaken in the literature review. Conclusions may state what needs to happen or explain cause and effect issues. Your recommendations should come from your data, and should not be too ambitious or suggest radical change in government or industry direction that is unlikely to materialise. To recommend another study may be appropriate. Appendices should include all material that is not necessary to meet study objectives. At the end of your study you should undertake your own appraisal, and recognise the limitations of what you have done, including judgements about reliability and validity. You will need to undertake a lot of early reading and make sure you write as you go.

# References

ABECAS (2005) Guidance to higher education institutions: removing barriers and anticipating reasonable adjustments for disabled students in built environment degree programmes. Farrell, P. & Middlemass, R. The University of Bolton. Available at http://www.cebe.heacademy.ac.uk/news/abecas/index.php

Audit Commission (2007) Local quality of life indicators—supporting local communities to become sustainable. http://www.auditcommission.gov.uk/nationalstudies/localgov/Pages/localqualityoflifeindicators.aspx

BCIS (2010) The Building Cost Information Service. Service.bcis.co.uk

Bell, J. (2005) *Doing Your Research Project: A Guide for First Time Researchers in Education, Health And Science.* 4th Edition. Milton Keynes: Open University Press.

Bothamley, J. (1993) *Dictionary of Theories: More than 5000 Theories, Laws, and Hypotheses Described.* London: Gale Research International.

Brook, N. & Farrell, P. (2004) An investigation to determine whether the approach to partnering by clients and contractors influences the success of projects. *4th International Postgraduate Research Conference.* University of Salford. 1–2 April 2004, pp. 608–609.

Brundtland, G.H. (1987) Our Common Future. Report of the World Commission on Environment and Development. http://www.worldinbalance.net/intagreements/1987-brundtland.php

Bryman, A. & Cramer, D. (1997) *Quantitative Data Analysis with SPSS for Windows.* London: Routledge.

Bryman, A. & Cramer, D. (2005) *Quantitative Data Analysis with SPSS 12 and 13.* London: Routledge.

BS 5605 (1990) British Standard 'Recommendations for citing and referencing published material'. 2nd Edition.

Business (2010) Business balls. http://www.businessballs.com/freepdfmaterials/XY_Theory_Questionnaire_2pages.pdf

Charlton, B.G. & Andras, P. (2003) *The Modernization Imperative; A Systems Theory Account of Liberal Democratic Society.* Exeter: Imprint Academic. pp. 85.

Clarke, V.L. & Cresswell, J.W. (2008) *The Mixed Methods Reader.* London: Sage.

Clough, P. & Nutbrown, C. (2007) *A Student's Guide to Methodology.* London: Sage.

Cohen, M. & Holliday, L. (1996) *Practical Statistics for Students.* London: Chapman Publishing.

Constructing Excellence (2009) KPI launch 2009. http://www.cewales.org.uk/cew/wpcontent/uploads/2009-KPI-Launch-Cardiff.pdf

Constructing Excellence (2009) 'G4C Egan Report' http://www.constructingexcellence.org.uk/news/article.jsp?id=9865 [09 September 2009]

---

*Writing a Built Environment Dissertation: Practical Guidance and Examples.* Peter Farrell.
©2011 Peter Farrell. Published 2011 by Blackwell Publishing Ltd.

Creswell, J.W. (1997) *Qualitative Inquiry and Research Design: Choosing Among Five Traditions.* London: Sage.

Creswell, J.W. (2003) *Research Design: Qualitative, Quantitative, and Mixed Methods Approaches.* London: Sage.

Egan, J. (1998) Rethinking Construction. The Report of the Construction Task Force. London: DETR. TSO

Energy Saving Trust (2010) Climate change. http://www.energysavingtrust.org.uk/ClimateChange? gclid=CKG4qe3mpqACFQE8lAoddSeEZQ

Fellows, R. & Liu, A. (2008) *Research Methods for Construction.* 3rd edition. Wiley-Blackwell: Oxford.

Fenn, P. (1997) Rigour in research and peer review. *Construction Management and Economics,* 15, 383–385.

Fink, A. (1995) *The Survey Handbook. The Survey Kit.* Volume 1. London: Sage.

Hart, C. (2001) *Doing a Literature Search.* London: Sage.

Hays, W.L. (1988) *Statistics.* 4th Edition. London: Holt, Rinehart and Winston Inc.

Hinton, P.R. (2004) *Statistics Explained.* 2nd Edition. London: Routledge.

Holt, G. (1998) *A Guide to Successful Dissertation Study for Students of the Built Environment.* 2nd Edition. The University of Wolverhampton, UK.

HSE (2005) Health and Safety Induction for Smaller Construction Companies. The Health and Safety Executive. http://www.hse.gov.uk/construction/induction.pdf [09 September 2009]

HSE (n.d.) The work at height regulations; case studies. The Health and Safety Executive. http://www.hse.gov.uk/falls/downloads/4.pdf [04 March 2010]

Joyce, H. (2001) Adam Smith and the invisible hand. http://pass.maths.org/issue14/features/smith/index.html

Kinnear, P.R. & Gray, C.D. (1994) *SPSS for Windows Made Simple.* East Sussex: L. Erlbaum Associates Ltd.

Kinnear, P.R. & Gray, C.D. (2008) *SPSS 16 Made Simple.* East Sussex: L. Erlbaum Associates Ltd.

Langford, V. (1988) Stress, Satisfaction and Managers in the Construction Industry. *Occupational Psychologist,* 6, 30–32.

Latham, M. (1994) Constructing the Team. Final Report on the Joint Review of Procurement and Contractual Arrangements in the UK Construction Industry. July. London: HMSO.

Matthews, R. (1998) The great health hoax. Sunday Telegraph Review.13 September.

McCraken, G. (1988) *The Long Interview. Qualitative Research Methods.* London: Sage.

Microsoft (2007) Microsoft Office; Word. Microsoft Corporation. Available at MicrosoftStore.com

Miles, M.B. & Huberman, A.M. (1994) *Qualitative Data Analysis.* London: Sage.

Naoum, S.G. (2006) *Dissertation Research and Writing for Construction Students.* 2nd Edition. Oxford: Butterworth-Heinemann.

Office for National Statistics (2001) Count me in; Census 2001 England Household Forum. http://www.statistics.gov.uk/census2001/pdfs/engh1.pdf

Office for National Statistics (2009) Construction statistics annual. http://www.statistics.gov.uk/downloads/theme_commerce/CSA-2009/Opening-page.pdf

Oppenheim, A.N. (1992) *Questionnaire Design, Interviewing and Attitude Measurement.* 2nd Edition. London: Cassell.

Richardson, S. (2009) Morrell takes the stage with a warning to government. *Building,* 27 November, Issue 47, pp. 10–11.

Rudd, D. (2005) Unpublished handout University of Bolton; adapted from Brown, G & Atkins M. (1988) *Effective Teaching in Higher Education.* London: Routledge.

Seymour, D., Crook, D. & Rooke, J. (1997) The role of theory in construction management: reply to Runeson. *Construction Management and Economics,* 16, 109–112.

Siegel, S. & Castellan, N.J. (1988) *Nonparametric Statistics for the Behavioural Sciences.* 2nd Edition. London: McGraw Hill.

Silverman, D. (2001) *Interpreting Qualitative Data.* 2nd edition. London: Sage.

Somekh, B. & Lewin, C. (2005) *Research Methods in the Social Sciences.* London: Sage.

Stott, M. (2010) An investigation into quality issues in housing completions. Unpublished work in progress. MSc Construction Management. The University of Bolton.

*Sunday Times* (2009) The *Sunday Times* 100 Best Companies. http://business.timesonline.co.uk/tol/business/career_and_jobs/best_100_companies/best_100_tables/

Syque (2010) Success through understanding. http://syque.com/quality_tools/toolbook/Variation/Image375.gif 5th March 2010

Turnitin (2009) Turnitin original checking. https://www.submit.ac.uk/static_jisc/ac_uk_index.html

University of Bolton (2006) Code of Practice for Ethical Standards in Research Involving Human participants. http://www.bolton.ac.uk/Students/PoliciesProceduresRegulations/AllStudents/Research Ethics/Documents/CodeofPractice.pdf

University of Bolton (2010) Research Ethics Checklist. http://www.bolton.ac.uk/Students/PoliciesProceduresRegulations/AllStudents/ResearchEthics/Documents/CodeofPractice.pdf

# Bibliography

Blaikie, N. (2003) *Analysing Quantitative Data.* London: Sage.

Bulmer, M. (Ed.) (1984) *Sociological Research Methods: An Introduction.* 2nd Edition. New York: Macmillan.

Clegg, F. (1982) *Simple Statistics: A Course Book for the Social Sciences.* Cambridge: Cambridge University Press.

Greene, J. & Oliveira, M. (1982) *Learning to use Statistical Tests in Psychology: A Students Guide.* Milton Keynes: Open University Press.

Howard, K. & Sharp, J.A. (1983) *The Management of a Student Research Project.* London: Gower.

Litwin, M. (1995) *How to Measure Survey Reliability and Validity. The Survey Kit. Vol. 7.* London: Sage.

McCracken, G. (1988) *The Long Interview. Qualitative Research Methods.* London: Sage.

Miles, M.B. (1994) *Qualitative Data Analysis.* California: Sage.

Patton, M.Q. (2002) *Qualitative Research and Evaluation Methods.* 3rd Edition. Sage: London.

Ruddock, L. (1995) *Quantitative Methods for the Built Environment. Volume 1: Statistical Analysis.* Warrington: White Castle.

Ruddock, L. & Knight, A. (2008) *Advanced Research Methods in The Built Environment.* Oxford: Wiley-Blackwell.

Strauss, A. & Corbin, J. (1998) *Basics of Qualitative Research.* 2nd Edition. Sage: London

Taylor, J. (2007) An effective methodology to ensure quality dissertations. Available at http://cebe.cf.ac.uk/BPBN/casestudy/ulster_cdc2i.htm [Accessed 05.10.07]

Taylor, S. & Bogden, R. (1998) *Introduction to Qualitative Data Research Methods.* 3rd Edition. New York: Wiley.

*Writing a Built Environment Dissertation: Practical Guidance and Examples.* Peter Farrell.
©2011 Peter Farrell. Published 2011 by Blackwell Publishing Ltd.

# Appendices

List of appendices:

A  Research ethics checklist
B  Narrative of a problem
C  A review of theory and literature
D  Qualitative analysis
E  Using Excel for charts, descriptive tests and inferential tests
F  The standard normal distribution tables
G  Chi-square table
H  Mann–Whitney table, $p = 0.05$
I  Mann–Whitney table, $p = 0.01$
J  Wilcoxon table
K  Related $t$ test table
L  Spearman's $rho$ table
M  Pearson's $r$ table
N  $F$ distribution

*Writing a Built Environment Dissertation: Practical Guidance and Examples.* Peter Farrell.
©2011 Peter Farrell. Published 2011 by Blackwell Publishing Ltd.

# Appendix A: research ethics checklist

(Source: adapted from University of Bolton 2006.)

Answer 'yes' or 'no' to each question:

- Will the study involve participants who are particularly vulnerable or who may be unable to give informed consent (e.g. children, people with learning disabilities, emotional difficulties, problems with understanding and/or communication, your own students)?
- Will the study require the co-operation of a gatekeeper for initial access to the groups or individuals to be recruited (e.g. students at school, members of self-help group, residents of a nursing home)?
- Will deception be necessary, i.e. will participants take part without knowing the true purpose of the study or without their knowledge/consent at the time (e.g. covert observation of people in non-public places)?
- Will the study involve discussion of topics which the participants may find sensitive (e.g. sexual activity, own drug use)?
- Could the study induce psychological stress or anxiety or cause harm or negative consequences beyond the risks encountered in normal life?
- Will the study involve prolonged or repetitive testing?
- Will financial inducements (other than reasonable expenses and compensation for time) be offered to participants?
- Will participants' right to withdraw from the study at any time be withheld or not made explicit?
- Will participants' anonymity be compromised or their right to anonymity be withheld or information they give be identifiable as theirs?
- Might permission for the study need to be sought from the researcher's or from the participants' employer?
- Will the study involve recruitment of patients or staff through the NHS?
- If ANY of the above items are answered 'yes' you will need to describe more fully how you plan to deal with the ethical issues raised by your research. This does not mean that you cannot do the research, only that your proposal will need to be approved by a Research Ethics Committee or Sub-committee

# Appendix B: narrative of a problem

(Source: Yarsley 2008.)

Prospective clients have the role of selecting the method of procurement for the required works of construction. It is essential that the correct method of procurement is chosen due to their financial input and the numerous risks associated with the construction industry. Sawczuk (1996) states construction projects vary from simple to complex. However, they all have something in common; they can still go wrong. Unfortunately, projects within the construction industry often do go wrong and the industry has a bad image amongst clients and the public. This is exemplified in projects exceeding cost budgets, finishing beyond schedule and being engaged in long disputes (Godfrey 1996). Loosemore *et al.* (2006) discussed this further by stating the construction industry has prominent examples of poor service delivery and poor safety records that have created negative public perceptions. Clients and contractors often lose out financially when projects finish late, over budget or below the required standards of quality.

Egan (1998, p.7) states 'projects are widely seen as unpredictable in terms of delivery on time, within budget and to the standards of quality expected'. The industry is under-achieving in performance levels of delivering construction projects; this was made evident in several independent reviews of construction in the 1990s. In 1999, a benchmarking study of construction projects for 66 central government departments, revealed 75% of projects exceeded their budgets and two-thirds had exceeded their primary completion date (OGC 2007). Current data obtained from Constructing Excellence's Key Performance Indicators (KPIs) show 46% of projects were completed within budget and 58% delivered on time in 2007 (Building 2008). The data set shows improvements have been made; however, the results are still unsatisfactory and the industry is still under-performing. A major contributor to projects exceeding budget is the provision of insufficient detail and information from consultants at design stage. This often leads to variations to the contract. In addition to the contract sum, variations have to be paid for. This generally results in contractors being paid through the submission of dayworks for carrying out extra work. Contractors may also apply for an extension of time, resulting in projects finishing beyond their completion date. Client satisfaction level for contractors' and consultant's work is another important indicator in the performance levels of the industry, as they pay their fees. A British Property Federation survey in 1997 revealed over a third of major clients were dissatisfied with contractors' performance, in delivering projects of the required quality, to the specified budget and date and resolving defects (Egan 1998).

The construction industry in the UK employs around 2.1 million people in a multitude of roles, and is currently the country's largest industry (DTI 2008). It is one of the most dangerous; over 2800 people have died from injuries they received as a result of construction work in the last 25 years (HSE 2008a). Each year construction workers are killed, injured or made ill whilst carrying out their work. There were 77 fatal injuries to construction workers in 2006/2007; this accounted for 32% of all worker deaths in 2006/2007 (HSE 2008b). Often it is the negligence and incompetence of fellow employees that cause the deaths and injuries. The general public that have to commute near construction sites are also at risk. Major injuries to employees in 2006/2007 numbered 3711 and based on the Labour Force Survey non-fatal injuries in construction were 1600 per 100 000 workers in 2005/2006 (HSE 2008b). The data accounts for reported injuries only and therefore the true figures could be higher. The industry's rates of death, injury and illness are far too high compared with other industries, and many accidents could be avoided. Griffith and Howard (2001) state accidents have a financial impact on the industry costing millions of pounds annually. Egan (1998) furthers this by stating accidents can account for 3–6% of overall project costs.

Egan (1998) states the industry is under-achieving and needs to modernise. The performance levels of the construction industry are unsatisfactory. *Constructing the Team* (1994) by Latham and *Rethinking Construction* by Egan (1998) set improvement targets for the industry. Latham (1994) concludes that cost savings of up to 30% of construction costs are possible with a will to change (Godfrey 1996). Egan (1998) set seven improvement targets to be achieved yearly. They included: reduce capital cost and construction time by 10%, increase predictability by 20% and reduce accidents by 20%.

Provisional objective: To determine whether different levels of contractor involvement in Design and Build procurement influences project success.

## Appendix C: a review of theory and literature

To put the literature review in context, it is also preceded by a brief description of the problem. (Source: adapted from Rucklidge and Farrell 2006.)

### Introduction and the problem

It is acknowledged that a highly motivated workforce can make the difference between success and failure (Ghoshal 1995, quoted in Crainer and Dearlove 2001). Walker and Smithers (1996) support this view, adding that 'business performance is inextricably linked to attracting, retaining and developing motivated and committed individuals'.

There are many variables that will come into play during the lifetime of construction projects, a number of which could well be beyond the control of project teams. It is important, therefore, for effective managers to understand what factors they can directly influence. Managers are responsible for obtaining results and, consequently, they rely upon the basic principle of understanding the most important resource; people. Inefficient use of labour and waste of resources in construction is common place. Latham (1994) and Egan (1998), in their respective reports 'Constructing the Team' and 'Rethinking Construction' identified adversarial attitudes, wastage and inefficient use of labour, as major characteristics of the modern day construction industry.

The Employers' Skills Need Survey, undertaken by the CITB, surveyed nearly 500 construction companies across Great Britain. These companies were asked about workload and recruitment difficulties in October 2003. Nearly 70% said that they were experiencing recruitment difficulties, with craft trades and managers representing the greatest difficulties (CITB 2003). Arguably then, there is a severe shortage of labour within the construction industry. Further evidence for this labour shortage is highlighted in the Built Environment Professional Skills Survey 2003/2004 (CIC 2004). Here, 68% of the 927 companies surveyed from within the construction professional services sector had experienced difficulties in the past 12 months with the recruitment and retention of staff.

There is increasing expectation by both clients and industry for greater efficiency and predictability in time, cost and quality. Despite numerous improvements in construction technology and procurement routes, actual productivity is arguably below par, when compared with other industries. This assertion is supported by Egan (1998), who states that 'there is deep concern that the industry as a whole is under-achieving'. Coupled with the predominantly labour-intensive nature of the industry, this suggests that 'proper emphasis should be given to such matters as communications, participation and motivation' (Mansfield and Odeh 1991). Understanding what motivates employees individually and moreover collectively, would enable managers to best allocate resources. Construction managers could be accused of marrying the concept of motivation to the use of incentives, and incentives with tangible rewards. The preconceived notion that motivation can be produced and sustained from the timely provision of tangible rewards is arguably short-sighted and highlights a lack of understanding and confused thinking in a complex area. Price (1992) acknowledges this point and argues that it is the transient nature of the construction industry that explains the trend towards financial incentives as the favoured method. Non-materialistic motivators, such as interpersonal relationships or career development, would be difficult to establish when short-term relationships are the norm.

Although many may argue that it is the nature and characteristics of the construction industry that have facilitated this practice for so long, the literature review provides a tantalising proposition that social class has a direct relationship with motivation. This, therefore, provides the catalyst for the study, that is, to see whether employees can be categorised into their social class and whether there are common trends and patterns in motivational needs or motivational levels across these groups. The ability of managers to make their workforce productive is fundamental to successful projects. Motivating employees to work in an efficient and productive manner is seen as one part of the 'traditional form' of management. The construction industry needs to concern itself with motivation at two different levels: to attracting the best people to fill the positions and, once in these positions, to motivating the people to perform the duties associated with them.

## The Literature

Motivation has been extensively reviewed in construction; therefore only some aspects of it are summarised here. Less interest has been taken in the concept of social class, and therefore a more extensive appraisal is presented.

### Motivation

The word 'motivation' was derived from the Latin word *movere*, to move. Hertzberg stated that motivation is a 'function of growth from getting intrinsic rewards out of interesting and challenging work'. Hollyforde and Whiddett (2002) define motivation as 'whatever the behaviour, the drive pushing or pulling a person to act in a particular way is motivation', and state that, 'most academics consider motivation to represent the drive behind human behaviour'. It has been suggested that effective management requires an understanding of what motivates a workforce. This understanding has changed substantially over the last century and is often believed to reflect the current labour situation. During the 1930s the threat of unemployment and the use of aggressive management achieved a form of 'productivity'. In later years there has been a reversal; financial rewards, over payments, incentive schemes, etc., resulting in a more persuasive form of management (Seeds 1999).

This change in management thinking is exemplified by the work of Fredrick W. Taylor, in the early part of the last century. Taylor's work on scientific management is often referred to as the traditional approach to motivation. Taylor's systems are based on employee pay; those who produced the best in quality and quantity received the best remuneration. It was believed these rewards would improve motivation and consequently, greater productivity and efficiency in the work place. However, a number of researchers and academics argue that:

'bonus, payment by results, financial incentives (or any synonymous method for systematically encouraging productivity) has been the grumbling appendix to industrial relations in the building industry, with arguments for and against its use, thoroughly rehearsed over some seventy years' (Hague 1985).

Taylor's work was put into question when Elton Mayo conducted the Hawthorn studies. Mayo selected six women from a factory. Mayo continually changed the women's working conditions, sometimes for the better and sometimes for the worse; always discussing and explaining the changes in advance. To his surprise, there was a significant improvement in production, irrespective of the change to working conditions. These findings were partially at odds with Taylor, who based his approach on the premise that workers are purely motivated by economic self-interest. Mayo's findings illustrate a fundamental concept: 'workplaces are social environments and within them, people are motivated by much more than economic self-interest' (AJA 2003). Mayo's understanding of the workplace as a social environment and of people, motivated by their role within this social environment, is clearly sound and potentially transferable to the construction industry.

Amongst others, Maslow and Hertzberg regard 'motivation' as being a highly complex issue and would suggest that human motivation is based on a series of needs and wants. Maslow and Bennis (1998) propose that these needs can vary from the basic necessities for life, such as oxygen, food and water to love, belonging and esteem. Maslow and Bennis could be open to criticism, in that they lend insufficient consideration to the effect of job and work-related variables. Price (1992) would say that 'a number of the early theories were primarily based on the individual's motivation, although job-related and work environments are not ignored. The main emphasis was placed on the individual's characteristic and the role played by personal needs in determining work behaviour'. Ruthankoon and Ogunlana (2003) would go further, in that Herzberg's theory, whilst widely known in management circles, has been criticised 'regarding its validity in work settings'.

Many authors, such as Ruthankoon and Ogunlana (2003), Gilbert and Walker (2001), Price (1992) and Hague (1985) use the established and well-known motivational theories of Maslow and Hertzberg amongst others, as the framework for their research. It was these theories that first recognised motivation as a series of needs and wants.

The characteristics of the construction industry (argued by some in Egan [1998], to be different to other industries, particularly manufacturing) may well limit the viability of recognised research undertaken in other sectors, as a basis for comparison. Ruthankoon and Ogunlana (2003) emphasise that 'construction is an industry with unique characteristics which may have special effects on employee motivation', and Price (1992) highlighted that the transient nature of the construction industry with short-term projects is a major factor. Arguably, there are many preconceived ideas about motivation, based on Fredrick W. Taylor's systems, in the early part of the last century. Hague (1985) states 'it is tempting to conclude that managers are motivated while manual workers need bonus payments'. In contradiction, McFillen and Maloney (1988) believe that each worker is unique, which may consequently limit the potential to identify trends within the industry.

### Social class

Davis and Moore (1945) link the two variables of social class and motivation. They believe that the inequalities created by social class in fact compel people to better themselves, ensuring that the most valued positions in society are filled by those most qualified and competent. In a modern society, some academics argue that it is no longer appropriate, justified or indeed, ethical, to categorise people into social class. Holt and Turner, quoted in Crompton (1993) assert that 'class is an increasingly redundant issue'. The continuing debate on social class and its value or otherwise to society is testimony to the need for further research into social classification.

At its most basic, social class is based on grouping people that have a similar social status. Liversey (2004) believes that social class is a widespread concept with many dimensions. The three primary dimensions are: (a) economic—i.e. wealth, income and occupation, (b) political—i.e. status and power, and (c) cultural—i.e. lifestyle, value beliefs, levels of education and so forth. The economic dimension is commonly accepted as the most important. This is largely due to the close relationship between economic position, social status and power. Liversey (2004) 'believes it is important to define social class because it is objectively linked to an individual's "life chances" '. Once defined, Liversey (2004) believes that it is then necessary to 'operationalise' social class; this is to find a method to gauge class, an indicator of class. This gauge could then be used as a basis for measurement and testing. Historically, this was done on the basis of occupation.

Rose (2004) states that the 'practice of officially classifying the British population according to occupation and industry began in 1851' and was undertaken for the purpose of mortality analysis. However, it was not until 1911 that a summary of occupations, designed to represent 'social grades' was introduced by the Registrar General's Annual Report. These were later referred to as 'social classes', originally named the Registrar General's Social Classes (RGSC) and then re-named social class based on occupation in 1990. Therefore, the Registrar-General's class scheme was based on the premise that society is formed by a hierarchy of occupations. This hierarchy is divided into five basic social classes: professional, managerial, skilled occupations, partly-skilled occupations and unskilled occupations. These classes were recognised by the Office of Population Census and Surveys (OPCS) and were described, from 1921 to 1971, 'as an ordinal classification of occupations according to their reputed 'standing within the community' (Rose 2004).

In the UK 'occupations were placed in social classes on the basis of judgements made by the Registrar General's staff and various other experts with whom they consulted, and not in accordance with any coherent body of social theory' (Rose 2004). Maguire (2004) states that the Registrar General's scale contains absurdities 'like equating small tenant farmers and major land owners' and then argues that 'the level of skill at work may not be the best way of measuring access to social resources'.

It could be argued that the concept of social class is wide-ranging. Liversey (2004) believes 'it is extremely difficult to operationalise since it involves a large number of variables such as the relationship between income, wealth, power, status, gender and age'. Modern attitudes and values may also limit the relevance of social classification in today's society. Crompton (1993) highlights a number of arguments to the effect that 'the idea of class is out of date and of declining significance' and that 'transformations of work and the structure of employment have blurred established class boundaries', whilst Breen and Rottman (1995) highlight the point that in Britain it is 'differences in consumption, rather than production, which are nowadays central to the formation of interest groups in society'. Liversey (2004) goes on to state that social class 'is not easy to define (although there may be certain observable indicators of a peoples' class, such as the way they talk, the way they dress and so forth)'.

Such criticism by academics and researchers may well have contributed to the National Statistics Socio-economic Classification (NS-SEC), introduced in 2001. NS-SEC replaces social classification based on occupation for official statistics and surveys. NS-SEC aims to assess employment relations and conditions; conceptually these are central to delineating the structure of socio-economic positions in modern societies and helping to explain variations in social behaviour and other social phenomena. The NS-SEC aims to differentiate positions within the labour markets and production units in terms of their typical 'employment relations' (NS-SEC 2004).

The positions of people in the labour market are determined by their level of income, security of income and prospects for improving level of income. Their position in the work environment is determined by the systems of authority and control at work, autonomy being a secondary aspect. The NS-SEC categories thus distinguish different positions (not people) as defined by social relationships in the workplace, i.e. by how employees are regulated by employers through employment contracts (NS-SEC 2004).

Stratification describes the different 'layers' that exist in society and is defined by Davis and Moore (1945) as 'unequal rights and prerequisites of different positions in society'. They propose that the 'stratification system provides for the appropriate motivations for the people to seek to fill certain role positions and desire to perform the appropriate tasks required of that position adequately'. Davis and Moore (1945) also support social stratification as an incentive for social betterment. They attribute more importance to particular jobs, for which they believe only a minority within society are able to perform. Tumin (1953) disagrees, on the basis that margins of society are not given the opportunity to discover their talent. The rewards used to motivate people and the hierarchical distribution of those awards are part of the social order and give rise to stratification. It could be argued that in addition to the motivational benefits of the rewards on the individual, the inequality created by stratification also has a motivational effect on society.

Wrong (1999) states that since the publication of the Davis-Moore theory, 'historical events and trends in social theory have increased the credibility of their work' whilst adding that 'their emphasis on rewards as incentives for individuals' self-recruitment to occupational roles was also right'. Wrong (1999) did, however, criticise their failure 'to mention that motives other than the desire for rewards may influence individual choice'.

Arising from the literature review the following two hypotheses are established: (a) social class influences motivation to work; where social class is the independent variable (IV) and motivation to work is the dependent variable (DV1), and (b) social class influences the impact of known motivational factors (where social class is the independent variable (IV) and the impact of known motivational factors is the dependent variable (DV2).

# References

AJA (2003) *Hertzberg, Maslow, Adams and Extrinsic Motivators.* Online. Available: http://www.aja4hr.com/management/hertzberg_motivation.shtml [10 September 2004]

Breen, R. & Rottman, D. (1995) *Class Stratification.* Hemel Hempstead: Harvester Wheatsheaf.

CIC (2004) *The Built Environment Professional Service Skills Survey 2003/2004.* Online. Available: http://www.cic.org.uk/services/SkillsSurvey2004%20Report.pdf [27 January 2006]

CITB (2003) *Employers' Skills Needs Survey'.* Online. Available: http://www.citb-construc-tionskills.co.uk/pdf/research/2003_skills_need_survey.pdf [13 January 2005]

Crainer, S. & Dearlove, D. (2001) *Financial Times Handbook of Management.* London: Prentice Hall.

Crompton, R. (1993) *Class and Stratification.* Oxford: Polity Press.

Davis, K. & Moore, W.E. (1945) Some principles of stratification. *American Sociological Review,* 10, 242–249.

Egan, J. (1998) *Rethinking Construction.* Online. Available: http://www.dti.gov.uk/construc-tion/rethink/repory [27 January 2006].

Gilbert, G.L. & Walker, D.H.T. (2001) Motivation of Australian white-collar construction employees: a gender issue? *Engineering, Construction and Architectural Management,* 88, 59–66.

Hague, D.J. (1985) Incentives and motivation in the construction industry. *Construction Management and Economics,* 3, 163–170.

Hollyforde, S. & Whiddett, S. (2002) *The Motivation Handbook.* London: Chartered Institute of Personnel Development.

Latham, M. (1994) *Constructing the Team.* Final Report of the Government/Industry Review of Procurement and Contractual Arrangements in the UK Construction Industry. London: HMSO.

Liversey, C. (2004) *Unit S3a: Measuring Social Class.* Online. Available: http:www.sociology.org.uk [28 November 2004]

Maguire, K. (2004) *Social Class.* Online. Available: http://www.medgraphics.cam.ac.uk/shield.html [12 November 2004]

Mansfield, N.R. & Odeh, N.S. (1991) Issues affecting motivation on construction projects. *International Journal of Project Management,* 9, 93–98.

Maslow, A. & Bennis, W. (1998) *Maslow on Management.* New York: John Wiley and Sons.

McFillen, J.M. & Maloney, W.F. (1988) New answers and new questions in construction worker motivation. *Construction Management and Economics,* 6, 35–48.

NS-SEC (2004) The National Statistics Socio-economic Classification User Manual, Version. 1.2. London: Office for National Statistics.

Price, A.D.F. (1992) Construction operative motivation and productivity. *Building Research and Information,* 20, 185–9.

Rose, D. (2004) *Social Classifications in the UK.* Online. Available: http://www.soc.surrey.ac.uk/sru/SRU9.html [15 November 2004]

Rucklidge, P. & Farrell, P. (2006) Social classification of construction employees influencing motivation to work. 6th International Postgraduate Research Conference in the Built and Human Environment. The University of Salford, 6–7 April at Delft University, Netherlands.

Ruthankoon, R. & Ogunlana, S. (2003) Testing Herzberg's two-factor theory in the Thai construction industry. *Engineering, Construction and Architectural Management,* 10, 333–341.

Seeds, R. (1999) *Motivation.* Unpublished lecture material. Bolton: Department of the Built Environment, University of Bolton.

Tumin, M. (1953) Some principles of stratification: a critical review. *American Sociological Review*, 18, 378–394.

Walker, D.H.T. & Smithers, G.L. (1996) Motivation and demotivation in the construction industry. *Chartered Building Professional*. May 17–18.

Wrong, D. (1999) Inequality and the division of labour. *Archieves Europeenes De Sociologie*, 40, 233–256.

# Appendix D: qualitative analysis

Files in this appendix:

(1)   Research objectives
(2)   Interview questions and prompts
(3)   Verbatim transcripts of interviews first copy; separate files for each interview, saved as file 3a, 3b, etc.
(7)   Verbatim transcripts of interviews third copy; originating from file 4, person A only
(8)   Verbatim transcripts of interviews fourth copy; for one of 17 labels only
(9)   A new file, comprising tables
(10)  The final narrative

## File 1: research objectives

Objective 1: to determine whether the propensity to put completions before quality influences profit within the private housebuilding sector (PHS)
Objective 2: to determine whether the propensity to put completions before quality is recognised by commentators within the current body of literature
Objective 3: to determine whether the propensity to put completions before quality is recognised by practitioners within the PHS
Objective 4: to determine whether the propensity to put completions before quality varies at different times of the year

## File 2: interview questions and prompts

Q1. What is the approximate proportion of subcontractors and directly employed labour currently working on this development?
Q1a. What are the differences between these two types of employee?

Q2. How would you describe your working relationship with your workforce?
Q2a. Under what circumstances are tradespeople pressured to work within unrealistic timescales?
Q2b. What are the implications for quality?
Q2c. How do tradespeople react to such pressure?
Q2d. How do subcontractors ensure that satisfactory quality is achieved?
Q2e. What is your policy regarding subcontractors' meetings?

Q3. How would you describe your working relationship with your contracts manager and directors?
Q3a. In what ways do these relationships change?
Q3b. What conflicts of interests are there when achieving completions?
Q3c. How does this issue affect your bonus payments?

Q4. In what way do you believe your build programme influences your ability to build houses to an adequate standard of quality?
Q4a. What additional help is available to you if needed?

Q5. Based on your experience, what would you say are the main reasons for sometimes being unable to build houses to an adequate standard of quality?
Q5a. Explain any reasons for feeling compelled to comply?

Q6. Describe how the build quality is controlled by this company?
Q6a. What causes some inspections to be missed out?
Q6b. How do you ensure there is sufficient time to rectify defects before the final check by the building inspector?
Q6c. Describe how thoroughly the building inspector checks a property during the final inspection?
Q6d. Explain your company's formal procedure for resolving defects?
Q6e. For what reasons is the system not always followed?
Q6e. What causes some defects to be rectified after occupation?

Q7. How would you compare customer satisfaction within the private housing sector to, say, 5 years ago?
Q7a. What effect do you think customer expectations and their perceptions of housing quality have on customer satisfaction?
Q7b. Do you feel customers tend to report more defects than, say, 5 years ago?
Q7c. At what stages can customers view their property?

Q8. To what extent do customers exhibit a willingness to recommend this company to family or friends?
Q8a. What about negative word-of-mouth referrals?
Q8b. How would you say this issue compares to, say, 5 years ago?

Q9. What awards for quality have you or any other site managers at this company been nominated for or received?
Q9a. What differences are there in how award-winning sites are run compared with the rest?

Q10. Finally, would you say that the opinions you have voiced in answer to these questions are commonly held by other site managers currently working for this company?
   End of interview.


## File 3: verbatim transcripts of interviews first copy

*File 3a: person A*

Q1. What is the approximate proportion of subcontractors and directly employed labour currently working on this development?

A-Ans1. Well the forklift driver's cards-in and I also have two bricklaying gangs who are cards-in but apart from that everyone else is a subbie. Even the labourers we use are all from an agency.

Q1a. What are the differences between these two types of employee?

A-Ans1a. I have to say that the standard of work I get off the cards-in brickies is noticeably better than the subcontracting brickies and they're also a lot tidier and more cooperative. You always find that directly employed forklift drivers are harder working, more reliable and more conscientious than agency drivers as well.

Q2. How would you describe your working relationship with your workforce?

A-Ans2. With most of them it's good, very good. There are a couple of the trades on here that I seem to have constant running battles with but I always manage to get what I want in the end so on the whole I'd say the working relationship's fine.

Q2a. Under what circumstances are tradespeople pressured to work within unrealistic timescales?

A-Ans2a. Sometimes, when you're really busy such as at year-end or half-year-end, you do pressure the subbies to work late or come in early and we definitely mither them to work weekends and that means full weekends too, all day Saturday and all day Sunday, not just a couple of hours on a Saturday morning.

Q2b. What are the consequences of this?

A-Ans2b. It becomes an absolute nuisance. You end up with things like the joiners first-fixing before the windows are fitted, which means the plot's not water-tight; the plumbers and sparks in together and pinching each others pre-drilled service holes in the joists; and the plasterers boarding the upstairs ceilings and one side of the stud walls while the building inspector's walking around the plot doing his pre-plaster inspection, which isn't ideal really. Then you might double up on your plasterers to get it plastered quicker but still have to put joiners second-fixing downstairs while they're finishing off skimming upstairs while the plumber's fitting his heating and sanitary wherever he can—it becomes madness at times, sheer madness.

Q2c. What are the implications for quality?

A-Ans2c. The implications for quality are the worst thing for me. I can deal with the long hours and trades having a 'barny' with one another but it's the bad effect on quality that really ****** me off at those times. A lot of the time I do think these situations occur simply because the construction director won't say no to the MD. I wish he'd just once say 'no, you can't have that plot in three weeks, it's impossible. You'll have to have it next month instead or it'll be a load of ****' or something like that.

Q2d. How do tradespeople react to such pressure?

A-Ans2d. The subbies themselves all basically do as they're told. The lads on site may moan about working weekends but they do get a bonus of about £25 per man, per day, for coming in, on top of whatever price-work they do, so at the end of the day they are earning more money for a few weeks. Ehm, as for their bosses, well, they moan too but at the end of the day they daren't refuse because they're **** scared of not getting any more work off us. I don't think any of them are bothered about quality really because if it is bad they can just blame us for making them rush it.

I'd have to say at times there are problems with health and safety too. Ehm, it's unavoidable really if you've got men working outside normal working hours. There's just bound to be times when someone ends up working on site without the benefit of a first aider present or even access to the first aid kit or toilets or washing facilities if the cabins are all locked up. Your painter can often end up as a lone worker too because they'll just lock themselves in a plot and stay to all hours to get it finished. I suppose it's quite dangerous practice when you stop and think about it.

Q2e. How do subcontractors ensure that satisfactory quality is achieved?

A-Ans2e. Well none of them use supervisors or foremen or anything like that, not really, no. I mean, if I have any issues with any of the lads on site regarding quality I'll ring their bosses and get them out to look at it with me and they always 'play ball', to be fair. Anyway, all you have to do is threaten not to pay 'em until they do come out and they'll soon turn up to sort it.

Q2f. What is your policy regarding subcontractors' meetings?

A-Ans2f. Well according to our company policy we are supposed to do them once a month and we have a pro forma on which to record the meeting and report it back to head office but no manager does them on a regular basis as far as I'm aware. I will always do one at the beginning of a new site as a way of introducing myself face-to-face to the subcontractors who I don't know or haven't used before. I find this initial meeting useful as it puts a face to a name and makes it easier in the future when you're dealing with them over the phone. Doing them each and every month though just isn't practical especially at crucial times of the year when you're mad busy because you just don't have the time and neither have they really. I mean you're talking to all the subbies on a daily basis anyway because the job's moving so fast so most of the managers don't see the point of them really and the contracts manager and construction director only seem to mention them when things are quiet and they have time to 'nit pick' over things—just like they do with the health and safety policy.

Q3. How would you describe your working relationship with your contracts manager and directors?

A-Ans3. Good, very good. We have a good understanding. I know what they expect and I've got high expectations, as have they, so, ehm, as long as we work together, we do a good job.

Q3a. In what ways do these relationships change?

A-Ans3a. I wouldn't say the relationship ever becomes tense or bitter as such, ehm, not really. I mean it can be difficult when you've got a lot of plots, a lot of completions in one particular month, ehm, but as long as we all work together and pull together, the help is already there so all you've got to do is ask for it.

Q3b. What conflicts of interests are there when achieving completions?

A-Ans3b. I think that naturally there is some, there would be a small amount of conflict there wouldn't there? Ehm, with them pushing to get profit by achieving the sales completions, which obviously increases the chances of them getting full bonuses. I wouldn't say it was unhealthy though, to have that.

Q3c. How does this issue affect your bonus payments?

A-Ans3c. Well my bonus is based on my NHBC items, reportable items that you get when the building inspector records defects he's spotted so it can affect how much bonus I get if they force me to hand over a plot with loads of items on it instead of giving me extra time to get them sorted.

Q4. In what way do you believe your build programme influences your ability to build houses to an adequate standard of quality?

A-Ans4. Overall I would have to say we are given enough time in the build programme to build a good house because quality seems to come before productivity at this company. Occasionally you are given less time than others depending on the plot and upon how and when it's been sold during a particular month and when the completion needs to be in for. Time pressures can be more severe on one plot than it could be for another but, you know, they're nearly always achievable. It's often last minute sales and stuff like that that cause the problems with time constraints. If we sell a plot and it has to be in within 4 to 5 weeks and it's achievable, we will certainly do our best to achieve it. They don't usually give you impossible scenarios. We can usually achieve what were asked to achieve.

I mean I have had situations in the past where I've been extremely rushed and where, say, a 16-week build, we've had to try and achieve it in 12 weeks. Well naturally, you know, I've not been happy with that, with having to do that, because with them extra 4 or 4 weeks, you can get things more pristine. The funny thing is though, when they do let you stick to the programme the pressures are probably greater because the expectations are higher because you've been given more time. Motivation is a key in this. You need to be motivated and ensure that you stay motivated, even though you might have a bit more time to complete it. You see, the overall expectation is far greater because you've had more time and once you've handed over the plot complete, they expect it to be to a really good standard, with greater emphasis on quality.

Q4a. What additional help is available to you if needed?

A-Ans4a. Probably the biggest negative aspect of working here, compared to other house builders I've worked for, is that cost controls are extremely tight. Everything you order is scrutinised and anything extra you ask for is questioned; even something like an additional broom for sweeping the welfare cabins out with. It becomes really frustrating and annoying at times, the way they 'penny-pinch', especially if they're asking you to build a plot in 30% less time than normal but they don't want to pay for temporary lighting even though we haven't got electric to the plot yet. They expect miracles on a shoestring budget but we usually get there in the end, which of course means that the situation never changes. The more often you do it that way, the more often they want it that way.

Q5. Based on your experience, what would you say are the main reasons for sometimes being unable to build houses to an adequate standard of quality?

A-Ans5. I haven't really experienced that within this company, no. With other companies I've worked for, yeah, but not here really, not with the current structure we've got. I think my finished units are always very good. We do have various stages and various viewings by both the contracts manager and construction director. We all look at it, you know, at various stages of the actual finishing stages and by the time we come to actually handing it over, I would say my standard's quite good.

You do get times when you want another day or two to finish a plot off but you're told no because they need it by that date to get the money in on time for the year-end deadline and sometimes it is difficult but the way I look at it, if it has to be in for that month, it has to be in. If it means putting a few extra hours in to achieve that, well we have to do it. If, you know, nine times out of ten it can be done, then yes, we certainly try our best to achieve it, although sometimes it's extremely difficult.

Q5a. Explain any reasons for feeling compelled to comply?

A-Ans5a. You do sort of feel compelled to do it, basically because it's the worry about losing your job and we've had that in the recent past. The demand can be so high that, ehm, that we would be rushed because of demand. Besides that, if we didn't co-operate and we didn't cope with demand, then we would either not get our bonus or our bonus would be extremely reduced.

Q6. Describe how the build quality is controlled by this company?

A-Ans6. It's down to the site manager mainly. The contracts manager will have a bob round about once a week when he comes to site but apart from that it's down to me. I do all the inspections really. We don't have a quality control manager or anything like that.

Q6a. What causes some inspections to be missed out?

A-Ans6a. I don't believe I ever miss any of the required inspections out. Like I mentioned earlier, our bonuses, the site managers' bonus that is, are only based on the amount of reportable items you get so it's in your own interest to thoroughly check plots before key-stage inspections by the NHBC inspector, otherwise you may end up with no bonus.

Q6b. How do you ensure there is sufficient time to rectify defects before the final check by the building inspector?

A-Ans6b. During busy periods like at year-end when you've got some tight plots because they've been thrown on you at the last minute, you might not get round all the plots as often as you'd like or spend as much time in each one as you normally would but you still make sure you do your checks, even if it means you being on site at six o'clock in the morning. Having said that, I've had many occasions where the NHBC (National House Building Council) inspector has turned up to do a CML while I'm half way through doing my final checks. You just have to keep your fingers crossed that you haven't overlooked any major defects when that happens.

Q6c. Describe how thoroughly the building inspector checks a property during the final inspection?

A-Ans6c. It really depends on which inspector you get as to how thorough he is. Some can be real pernickety ******** and others quite lax. My inspector on here is a bit of both; he can be okay one day and a complete 'jobsworth' the next. At times like year-end it also depends on how many visits they have on. With a lot of builders having they're year-end around the same time, if they're all wanting CMLs on the same day then he obviously won't have as much time to spend in each one so that can work to your advantage.

Q6d. Explain your company's formal procedure for resolving defects?

A-Ans6d. We have a procedure set out in the quality management books. What's supposed to happen is that a few days before the CML is done, the site manager and the sales negotiator 'snag' the plot and the site manager sorts the snags before the CML. After the CML is done, the sales people make an appointment with the client to come and view the property in what we call a 'home demo' but in the book it's called a 'Pre-Handover Viewing' and if the client picks up any other snags, the site manager is supposed to try his best to get them done before the client moves in. At the actual handover of the property to the client, which the book calls a 'Key Release', they're again asked to note down any defects but also to sign off where issues have been resolved. Then seven days after they have physically moved in, the site manager does another inspection with the client and once they've signed off the book to say he's done them the book goes into the office and the customers then have to contact head office direct with any problems.

Q6e. For what reasons is the system not always followed?

A-Ans6e. I say that's what's supposed to happen because it very rarely pans out that way. Firstly, the sales people—I won't say sales women because we do have a man as a sales negotiator—the sales people are all 'bone idle', especially when it comes to filling out the 'Quality Management' books. You see, it's their responsibility to do them. They have the books from day one and it stays with them until I do my seven-day call. I can't remember the last time I did a pre-CML check with a sales negotiator. I do do them of course but I do them on my own. The other thing is that, especially at year-end, you often have the scenario where clients are moving in the same day you get the CML or the house was only actually completed on the day of the CML so you obviously can't follow the procedures in the book. It's all a good idea in principle but I think it's mainly just a token gesture to try and make the company look efficient.

Q6f. What causes some defects to be rectified after occupation?

A-Ans6f. Some defects often have to be rectified after completion but it's nearly always due to time limitations. I don't think it's ever about the site manager just not trying hard enough. The other thing is that all house builders work within tolerances so there are certain things we don't have to attend to but the clients often think we do. If it's something that's only half-an-hour of a job, such as a painter touching up after they've had carpets fitted, then you'll often just do it as sort of a gesture of good will to try and get the customer on your side and establish a good relationship with them from the start. I would say we probably do around 30% of the snags after completion.

Q7. How would you compare customer satisfaction within the private housing sector to, say, 5 years ago?

A-Ans7. I think that house buyers are more demanding of good quality than in the past, yes, very much so, especially with the Internet and programmes on TV and in the general media, it's ehm, it is becoming more difficult as time goes by. I mean, the media in general over the past few years has promoted a bad image and people these days are expecting far more for what they're actually paying out, whereas in the past they were just, you know, happy to purchase a plot for instance and just go along with it but nowadays, they just, they want the best for what they're paying for.

Q7a. What effect do you think customer expectations and their perceptions of housing quality have on customer satisfaction?

A-Ans7a. I think the quality of the actual build has remained the same, if not improved; it's the expectations of the actual client, the purchaser that has risen. Their expectations have got far greater because they want value for money and they want what they want and they seem to know a lot more about what they want through the use of the Internet and the media. Ehm, I think our attitudes have remained the same and if the job, if the build is actually built to the way it should be along the process of the NHBC inspector looking at it at the various key stages and we all know as site managers what the company expects, then I'd have to say that quality has probably improved.

Q7b. Do you feel customers tend to report more defects than, say, 5 years ago?

A-Ans7b. If the defect's there to be reported, then the customer reports it, so yes, I suppose they do tend to report more defects now than a few years ago.

Q7c. At what stages can customers view their property?

A-Ans7c. The home demo is usually the first chance the customers get to see their property. It's the best time really; to see it when it's completely finished because it gives them a better impression of what they've actually bought. You sometimes get the odd one wanting to view the plot at first-fix or when it's plastered but site managers discourage this because clients end up walking round the house pointing at all sorts saying 'this isn't finished' or 'that isn't finished' and you feel like saying 'I know it's not finished you ****** fool, we're still building the ******* thing'.

Q8. To what extent do customers exhibit a willingness to recommend this company to family or friends?

A-Ans8. Ehm, not really sure about that one. We sometimes get clients giving us bottles of wine and stuff as a thank you if they're really happy with their new home and some do send letters in to head office praising the site staff but apart from that I don't really know. We've actually had relatives of people who work in our head office buy houses off us but that's because they think they'll get a good deal more than anything.

Q8a. What about negative word-of-mouth referrals?

A-Ans 8a. You only really get to hear from the ones who threaten to go to the press or threaten to tell everyone they know not to buy a ************ house but I don't think that's changed much over the years.

Q8b. How would you say this issue compares with, say, 5 years ago?

A-Ans 8b. Ehm, they used to reckon one in every 10 home buyers would be a 'customer from hell' so judging from that I don't think we get that many really bad ones but you never know what people say to their family and friends, do you?

Q9. What awards for quality have you or any other site managers at this company been nominated for or received?

A-Ans9. I was nominated this year but I haven't heard anything of that yet and as far as I know at least two other site managers have been nominated for 'Pride in the Job' awards.

Q9a. What differences are there in how award-winning sites are run compared with the rest?

A-Ans9a. For some companies I definitely know they are run differently but at this company, no, and that's mainly due to the cost constraints. We don't have a great deal of money to spend on the likes of fancy sales and site compound areas and other presentation ******** like other companies do. We do still try to achieve the standard of quality that would enable us to win an award but it's not really possible to achieve the overall standard that the NHBC is looking for when it comes to awards without spending loads of money 'tarting' the site up. Besides, everyone in the game knows that the biggest house builders take the **** bosses on golfing weekends and trips abroad and what have you so a lot of it's plain ******* really and little to do with the way they actually build the majority of their houses.

Q10. Finally, would you say that the opinions you have voiced in answer to these questions are commonly held by other site managers currently working for this company?

A-Ans10. Yes, definitely. I would say all of them. I know all the managers here personally and I can't think of any one of them, off the top of my head, who would disagree with anything I've said.

    End of interview.

*File 3b: person B*

Q1. What would you say is the approximate proportion of subcontractors and directly employed labour at this company?

B-Ans1. They're all sub-contractors, all 'subbies'.

Q2. How would you describe your working relationship with your workforce?

B-Ans2. In the main, I would say it was quite good.

Q2a. Under what circumstances are tradespeople pressured to work within unrealistic timescales?

B-Ans2a. It happens when they shorten the build programmes, that kind of thing. It just becomes more vertical all the time, the build programme and they end up having to work Saturdays, Sundays, Bank Holidays, seven or eight o'clock at night; all that business. It's basically during the run up to year-end from the summer up to September. The half-year-end is a similar scenario but not quite as much pressure. In between half-year-end and year-end it's more or less a steady pace.

Q2b. What are the consequences of this?

B-Ans2b. Well you feel under tremendous pressure because they want so many units in so you end up flooding it with men and you'll have every trade working on top of each other and coming up to September, year-end, you know that you'll start having to get the handovers coming in at the end of July and through August and if you're looking after 30 or 40 plots yourself and you're handing over six or seven a week, you're expected to go in and 'snag' them as well and you can't physically do everything. The amount they ask you for at year-end, you can't physically do it with the amount of hours in the day; there's never enough hours in the day. It's not that you can't do your job it's that you don't have enough hours to do it in.

On this site, when you've got people coming in to start work really early, you've got to come in yourself to open up because you can't trust the labourers to do it for you. You see, we are under obligation as part of the planning permission to employ the locals—well they are useless. So you're working with useless labourers and working in **** crowded conditions all the time.

To cope with being under that sort of pressure you have to work long hours and weekends as well. They never used to pay us for weekends but they've started paying us to soften the blow about having to keep giving up our weekends. I have times when I'm doing handovers and trying to 'snag' plots as well as do all my other jobs and I'll have some weeks where I'll turn in at half six in the morning, before everyone's got there, to start doing my work early, to be there to intercept subbies on their way in and what have you and then I'm going home and waking up in the middle of the night thinking about plots.

I find that it starts affecting my performance as a site manager as well. I won't be taking any breaks during the day and I suppose as the years go by you learn to manage the pressure better yourself but I'll have days when I won't have a lunch and you'll have the hierarchy or directors that come round and expect you to walk all the plots with them and that takes hours out of your day, keeps you from the work you should be doing.

Come the afternoon, sometimes your brain goes into 'meltdown' and your concentration levels just go, basically. In a morning you're 'on the ball' and whatnot but in the afternoon you'll be running out of steam, especially if you've been running up and down stairs in an apartment block. Then when I go home the missus is telling me I look gaunt and losing too much weight because you're just on the go so much all the time and you do start making mistakes yourself.

Q2c. What are the implications for quality?

B-Ans2c. Well the subbies, it kills the subbies and they end up pulling practically anyone off the street because they end up short of men. For example, they could have plasterers in they don't know or joiners they've never used before because they're under that much pressure to supply the men and they could mess the plot up. You don't have your regular subbie in all the time so the standard of workmanship ends up suffering.

If you've got plasterers still working in a house and you have to send painters in because of the lack of time, how's he going to have a chance of doing a decent job when he goes in? The joiners can't always keep up with the programme because he can't supply enough men and you end up with poor workmanship running all the way through. You can't physically, I can't get, there comes to a point where you can't throw a subbie off a job. If the work is poor and you aren't rushed you could just say 'I don't want that man anymore' and the job could hang on a day or two 'til you got the right man in but when you're rushed you just have to put up with it because otherwise you know you won't get the handovers. So, if I had a set of boarders who were absolutely useless, I'd just have to let them carry on boarding, what else could I do? If I throw them out I just wouldn't get the handover. All in all it becomes a nightmare. They all end up working on top of each other, over crowded, out of sequence, working late, often unsupervised, poor lighting because you often don't get services on 'til the death – a nightmare. I mean you'll have all the support to set up temporary lighting but just trying to maintain the lighting. . .if you've got 'festoon' lighting in and people are banging and bulbs are going. . . you can spend all day ******* around with festoon lighting in an apartment block and you're not getting anything else done. The bulbs gone in this box and these bulbs have gone here and is health and safety gonna come on and see there's a bulb missing and the cage has fell off and someone's just pulled a lead out of the transformer – you're into all that.

I mean there is always someone there who's a first aider. I'd have to say that on that one they're insistent that there will be a first aider there but supervision wise, I couldn't. If I have that many units and it's that busy and I have to supervise what all the trades are up to so, you know, I'll go into where someone has messed up a plot and everyone's blaming each other and there are so many trades in there that it just gets left a tip, which can obviously be dangerous health and safety wise.

When it's booming, like before this recession, the trades know they can go down the road and get another job so once you put them under a certain amount of pressure they'll just leave the job and **** off because they know there's another job to go to down the road. Now all this 'credit crunch' has come into being and work's become scarcer but we're still quite busy so I can hold the high ground and say 'well if you don't do it my way, go and I'll get someone else to do it'.

Q2d. How do tradespeople react to such pressure?

B-Ans2d. You do end up putting the subcontractors under a lot of pressure, i.e. their supervisors and I'll be, like, kicking his head in of a day and once the pressure starts getting to him, he starts forgetting things as well and making mistakes. I mean it definitely reduces morale and causes friction between you and your subbies because I'll end up talking to subbies like, ehm, well, putting them under pressure. Everyone is under massive pressure from, I suppose, the MD of the subcontractor down to his contracts manager and then he passes it onto the supervisor who takes it out on his lads.

They might put up some resistance at first but not in the end because they're obviously being told by their superiors and me just to get on with it. Their bosses might try to be awkward sometimes but you have them by the *****, the company has them by the ***** because the subcontractor is so reliant on us because they'll have so much work with us that they can't afford to say no. You can see it in some of their bosses' eyes when they come out to site; they can't say no.

Q2e. How do subcontractors ensure that satisfactory quality is achieved?

B-Ans2e. You will be looking at the quality as well yes but not so much to stop you throwing men at it. I mean, I might put a joiner in to second-fix a plot that will take him three days and then have to send the painter in after two days and then the joiner's final fixing while the painter's glossing out. Then, if I go to the painter before handover and tell him that his paintwork's a bit rough he'll just say 'what do you expect when I'm painting while the joiner's in and there's sawdust and all kinds of**** flying around' but you'll still sometimes force him to go back in and redo bits and threaten not to pay him if he doesn't.

Part of the subcontractors' package includes the responsibility to provide supervisors to monitor their lads on site. Now, every subbie is supposed to do that but it's a running battle to get them to stick to it. They will come out sometimes to check their lads' work but it's only 'once every blue moon'.

Q2f. What is your policy regarding subcontractors' meetings?

B-Ans2f. We do hold subcontractor meetings very regularly, depending on how busy we are. Sometimes during the really busy periods we might have one every day and all the subbies' bosses will have to attend—we even make the cleaner's boss come out (laughs **IntOb1**). During less busy times we'll probably only have them once a month or something but when you're busy it definitely helps to have them more regular because you can hammer out exactly what you need from each of them and sometimes I'll deliberately embarrass them by giving them a ********** in front of the other bosses if they've let me down in any way.

Q3. How would you describe your working relationship with your contracts manager and directors?

B-Anw3.Good and bad.

Q3a. In what ways do these relationships change?

B-Anw3a. In the times leading up to year-end and half-year-end, they'll be constantly on top of you all the time, screaming 'when's this happening, when's that happening'? They'll say 'you'd better go out there and grow a set of ********. Ring him up now, I want him here now. I want this many people in tomorrow' so there's a lot more pressure being caused by the handovers; by them wanting the amount of units they're looking for. The pressure goes from the MD to the construction director to the contracts manager and down to the site manager. Like some weeks I could be handing over ten units a week; in a week.

Q3b. What conflicts of interests are there when achieving completions?

B-Anw3b. The senior managers and directors bonuses are all profit-related, based on achieving profit forecasts, which obviously has a lot to do with why they push you so hard to get your completions in. Yes, I'd have to say that's why they want to get so many units over because that's what's going to affect their bonuses.

On a personal level I don't like having to hand over a plot to customers who are all exited about moving into their new home, sometimes supposedly their 'dream' home, when I know they're going to find all sorts wrong with it and be really disappointed with the standard. It's not a nice feeling.

Q3c. How does this issue affect your bonus payments?

B-Anw3c. To me it doesn't really make any difference or not whether a house goes over or not because I still get more or less the same bonus as long as I build it to the programme. My bonus is down to three factors: health and safety, building to programme and customer care. None of it is based on NHBC reportable items so as long as I get a CML it doesn't matter how many items I get. Obviously you don't want to get reportable items but as long as you don't get a 'red' item, which prevents the inspector from issuing you with the CML, then you can have 'green' items all over the place. It's just the 'red' items you mustn't get because that's a 'show-stopper'. So I have to say that we do get our bonuses, they do pay us our bonuses but they get their pound of flesh out of us; you know, the amount of work they put you through, the long hours and the sleepless nights but I'll always get my bonus, yeah.

Q3d. What additional help is available to you if needed?

B-Anw3d. More managers will sometimes come on board as the job goes on. On one phase you could have three assistants, two site managers, a project manager and a build manager so you could have five or six managers in total but you'd need that many. To be honest, a lot of the time when it's busy it's just as difficult running a smaller site with only one manager on it because there's still never enough time. More time is what you really need; not more men or equipment but they never give it to you.

Q4. In what way do you believe your build programme influences your ability to build houses to an adequate standard of quality?

B-Anw4. Everyone is looking at the programme and timescales all the time. The build programme when you look at it, the amount of weeks, yes at first looks enough but because it's so vertical it only takes one trade to start 'flagging' and going by the wayside to put the whole programme out of 'sync'. Then, you know, as happens, if the plumber starts to struggle on a number of plots and then the joiner, he's under pressure and he's got to try and get joiners to go back in there to do pipe-boxings after the painter's already been in there but you've had to do things out of sequence because the sanitary was late going in.

The trades are all on price so when they have to keep going back in again and again because of the order of works being mixed up, they're losing money and in their eyes doing work for nothing so they stop being bothered about doing a good job. Then you've got to get your painter back in again but then if you don't get the same painter back, you're getting some fuddy-duddy who just goes around snagging here and there, he's not the quickest, he's not the best and quality is going to suffer again.

There aren't any allowances made on your programmes for things like bad weather either? I mean, the past 2 years and all the rainfall we've had during the summers, I know no one could have foreseen that but there's absolutely no leeway whatsoever. I mean, there's not even anything in the programme for a scaffold strip and some strips can be massive, like on apartment blocks and they can hold you up by as much as a week. The programmes change all the time; you'll get issued with a new programme but it'll keep changing all the time. It starts off with a nice gradient and goes to looking almost vertical as they keep shortening it.

Thing is, you get to September, year-end, finish and then the programme is often delayed for the start of the next phase. Instead of starting back up again in October and starting digging footings you could have problems with land remediation and the QS' arguing over prices and so the programme starts late again. Instead of starting in October or November you wouldn't be starting proper again until January or February and in the meantime you'd have nothing much to do; it could be very quiet. So you can be weeks behind programme before you've even got on site so you know it's going to be unrealistic beforehand; you know what's coming.

Q5. Based on your experience, what would you say are the main reasons for sometimes being unable to build houses to an adequate standard of quality?

B-Anw5. A lot of time it's because of the workload. I know it's gonna come back and bite me because I haven't had enough time to do everything I need to do to get the quality to the right finish inside. You know that instead of just handing a house over and that's it and it's a nice job done and you can just walk away, you know that you're gonna have to go back in it; that someone from customer care is gonna go in and slaughter it and you're going to have to go back in.

I mean there's nothing wrong structurally wise, it's just the finishing quality that suffers. The defects are usually mainly in the finishing, in the finishing quality. For some plots you don't even have time to put a plaster patcher through them before you get them painted and then you'll have to go back in after handover and it often gets to the point where the painter is more or less doing a 're-dec' (redecoration) because they've had to gloss it without services on so there are runs all over the place because they were painted without any heating on, when the plot was still cold and damp and with other trades still working in there so you also end up with gritty paintwork; all because we were rushing the house through. Another common issue is having to send plaster patchers back in to cut holes in walls to find buried wires for the electricians because the plaster boarders have just boarded over them, you know, because they've been thrown in and told to do the job double-quick and they'll have been working at such a rate that they've screwed into loads of pipes too and then when you're testing the plumbing installations out, once you've got the services on, there'll be leaks everywhere but you can't be there all the time to monitor it because you're that busy. I suppose it's just music to their ears because they know they can just take the p**s and instead of taking the care to measure and cut holes and feed wires through for lights or alarms and such like, they just board them in.

Q5a. Explain any reasons for feeling compelled to comply?

B-Anw5a. You feel compelled to do these things because, well, there's an atmosphere of intimidation from the hierarchy who definitely try and intimidate site managers and pressurise them a lot, you know, to get the handovers, so you end up being scared about losing your job because if they want to get rid of you they can always find a way. Yeah, I do, I do worry about my job or worry about it affecting my chances of being promoted within the company.

Q6. Describe how the build quality is controlled by this company?

B-Anw6. We have a customer care department with a customer care manager but he only looks at the finished product. He doesn't inspect a house as its being built, that's my department, he only looks after the finished product.

The procedure is, or it's supposed to be, you get the CML, you hand it over to sales and once they've handed the paperwork into the office, within 3 days customer care will come out and go through the plot and 'snag' it. Then you have 5 working days to put right anything they've picked up and then you have to invite them back for a revisit to check it.

Q6a. What causes some inspections to be missed out?

B-Anw6a. Let's say on a decent standard house you can expect 15 to 20 snags, which is mainly bits of paintwork and plasterwork because that's what they mostly look at but when it's really busy you'll get some of them with, like, 150 snags on it and then you're having to go back in and you've got all these other units you've still got to knock out. You're having to go through a snag list that's done by someone else, cut it down into trades, ring the people up and fax them and it's taking up so much of your time that you're not looking through the next plots that are coming up to completion; your eye's off the ball. Really, they should get other people in to do that but they never do. You're expected to sort all that out so basically you're supposed to snag it yourself but sometimes the workload's so great that you can't and you know you're handing over **** and customer care are going to go in and shred it to bits. You know that's going to happen and that he's going to come back with a list of 100 or 200 items and yet you know that you've had to turn the house out like that because of the programme.

If you've got 10 plots with all them items it can take you hours and hours in a day to sort the jobs out from the snag lists from customer care. The way I do it is to photocopy loads of them for all the various trades because the quality is that poor that you'll have, like, 10 different subcontractors involved. Then I'll highlight the different jobs for each of them because you can't just give them a copy of the sheet and trust them to go through it conscientiously, you have to give it to them in, like, big highlighted things so they can see it and then you'll hand it to them but you won't have time to go back and check all that again. Then if customer care goes back in again and doesn't get it down to, say, five items, they'll fail it again and then there's a stewards enquiry with the hierarchy; 'why has it failed again?', 'why haven't you got it done?', 'well, I'm trying to do all this as well' and it just causes conflict and a bad atmosphere throughout the whole company.

So for some houses that you do, the procedure rarely works because you often have people moving in the same day as the CML and so the plot has to go over by default. So they just move in and then you have your fingers crossed as to how much they'll find. I mean, some punters will move in and they haven't got a clue and so they'll just go in and think its fine. Others would be all over it like a rash and then the emails to head office start. I'd say the proportion of houses that go over by default is probably between 5 and 10%, which isn't a massive amount as a percentage but still a lot of actual plots per year for the company as a whole.

Q6b. How do you ensure there is sufficient time to rectify defects before the final check by the building inspector?

B-Anw6b. Well the short answer to that is you can't. You just can't ensure there's enough time because there seldom is. No, not at all, nothing like sufficient time.

Q6c. Describe how thoroughly the building inspector checks a property during the final inspection?

B-Anw6c. To be honest we usually get away with murder with him although it depends on the inspector; some can be quite severe. It just depends but you can tell that they're under pressure to give you the handovers as well from our contracts managers and directors. I mean, one building inspector I had was just a thoroughly nice bloke and hardly ever gave me any items but his replacement was a complete pest. I suppose a lot depends on their individual personalities when it comes to how thorough they do their inspections.

Q6d. Explain your company's formal procedure for completion defects?

B-Anw6d. Before the client moves in they have a 'demo' tour of their home then after their demo they have up to 2 weeks to contact customer care with their own list of snags and then customer care go in and agree which items they'll rectify and which they won't because, you know, some clients might give you a never-ending list but you're only obliged to deal with the ones that fall outside of the tolerances that the NHBC allow because, you know, house construction is subject to tolerances. Then after they've done that list with the customer care manager, that then gets handed back down to site. I never really understand this properly though because once we've done all customer care snags, as far as I'm concerned we should be out of the equation because they're supposed to use their budget to go back in but it always comes back to site again, always ends up being at the site's cost.

Some of the things they pick up are just ****** ridiculous but it's all about job justification isn't it? I mean anyone can find fault with anything and when you go through so much s**t when you've got completions and then you get some ******** from customer care going in pointing out piddling little flaws that you need a magnifying glass to see; it just gets your back up.

Q6e. For what reasons is the system not always followed?

B-Anw6e. Well often a plot will go over by default. You can have a plot go over by default before customer care even go in to do their pre-handover inspection, if you get me, because if it's been right on the brink of year-end when you got it finished and then as soon as you got the CML the person will be moving in so customer care won't have a chance to go in, so, you know, it will be of a very poor quality. You never walk into one and say 'yeah, this is how I wanted to get it', you're always handing it over, looking at it, knowing that the quality is as poor as it is because you rushed it.

If you've got the CML and your customer care manager comes out and says 'right, you've got 200 defects here' but the customer's moving in the next day or even the same day, then the procedure goes out of the window and the customer just has to go in.

I'll give you a good example. We had a customer care guy who was a **** and I remember him going in this particular plot and recording 164 ******* items but it had to go over by default because the client needed to move in. When the client eventually did her snag list she only picked eight ******* snags up so I took the list back to the customer care manager and said 'go and ***** that ** your ******* ****'. I know it doesn't sound good but I end up wanting them to go over by default because I know the client will never pick up as many defects as what customer care will pick up. Plus, if it goes over by default then it's not my fault.

Q6g. What causes some defects to be rectified after occupation?

B-Anw6g. It doesn't seem to matter how many customer complaints you get, although you get a ********** for having too many, the company never stops and says 'we've got to revise the way we do things', no, it's all about the units because that's bonus-based for them you see. They're the guys sitting up in the office with the MD agreeing to get all these units because they know they're gonna get such a big bonus out of it. They're attitude is get the CML and sort the defects later. It's a tricky one to put a figure on but I'd say roughly about 10% of defects have to be rectified after completion, yeah, about 10%.

Q7. How would you compare customer satisfaction within the private housing sector with, say, 5 years ago?

B-Ans7. I'd say it was a lot less, yes, especially on the busy sites like this anyway, definitely less. I think the majority of houses are okay but quality is definitely getting worse overall. At the end of the day, the majority of the plots when the client moves in are reasonable but, well, you get your good ones and your bad ones. Sometimes you get bad plots and they'll all be condensed next to each other because they've come at the critical time leading up to year-end. That's where you're worst customers and complaints would come from.

Q7a. What effect do you think customer expectations and their perceptions of housing quality have on customer satisfaction?

B-Ans7a. I think a lot of them are aware of the TV programmes you see a lot on the telly now, these fly-on-the-wall documentaries. I think they're more 'clued-up' nowadays as to what they're getting. I've had someone come into a private house and start crying as soon as they walked in because the quality was that bad, the painting was terrible and they were literally crying. That's because they're expectations were shattered the moment they walked in and I couldn't blame them because some of them will often move in the same day they get the CML so you don't even have time to rectify anything before they do move in. You tend to remember the bad ones more than the good ones because the good ones you hand over, you tend to forget about them.

Q7b. Do you feel that customers tend to report more defects today than, say, 5 years ago?

B-Ans7b. I'd say so yeah, yeah. I would say we spend about 50% more time than we used to, rectifying defects to sub-standard houses than those we've been able to build to the quality we want.

Q7c. At what stages can customers view their property?

B-Ans7c. They can view it on completion of the build during the home demo with sales and customer care, once the CML is in.

Q8. To what extent do customers exhibit a willingness to recommend this company to family or friends?

B-Ans8. I don't know really. I never really deal with the customers apart from dealing with the snagging on the clients first list but even then they very often aren't in. A lot of the time you just get some keys off them and go in but generally it's the customer care department that has the most contact with the customers. They do the demo and they agree what we will attend to from the clients' first list.

Q9. What awards for quality have you or any other site managers at this company been nominated for or received?

B-Ans9. We actually won an award, a 'Pride in the Job' award, on the first phase of this site. I was only the assistant then so the site manager at the time was presented with the award but I also got a certificate as his assistant.

Q9a. What differences are there in how award-winning sites are run compared with the rest?

B-Ans9a. Well none really, not on here anyway. I mean we did have an advantage in the sense that it was only a small site then with 22 plots on it so with effectively two managers on it; it was pretty easy to run.

There was another site that won an award a few years ago though where they did throw money at it. You know, it had proper metal hoarding all the way round the perimeter instead of ***** old timber or ********* heras fence panels and a tarmac compound with flagged walkways and proper meeting rooms and all that. It also had some trees and nice landscaping around the compound too but that site did over 200 units in 6 months, which all sold so it made its money for the company in return.

Q10. Finally, would you say that the opinions you have voiced in answer to these questions are commonly held by other site managers currently working for this company?

B-Ans10. Definitely. All the site managers would say the same, all of them, all of the ones I've worked with here and that I know here. Yes.

*File 7: verbatim transcripts of interviews third copy; originating from file 4, person A only*

A-Ans1. Well the forklift driver's cards-in and I also have two bricklaying gangs who are cards-in but apart from that everyone else is a subbie. Even the labourers we use are all from an agency.

A-Ans1a. [I have to say that the standard of work I get off the cards-in brickies is noticeably better than the subcontracting brickies and they're also a lot tidier and more cooperative.] **QUAL3**

A-Ans1a. You always find that directly employed forklift drivers are harder working, more reliable and more conscientious than agency drivers as well.

A-Ans2. [With most of them it's good, very good. There are a couple of the trades on here that I seem to have constant running battles with but I always manage to get what I want in the end so on the whole I'd say the working relationship's fine.] **QUAL3**

A-Ans2a. [Sometimes, when you're really busy such as at year-end or half-year-end, you do pressure the subbies to work late or come in early and we definitely mither them to work weekends and that means full weekends too, all day Saturday and all day Sunday, not just a couple of hours on a Saturday morning.] T1/**PROP3**

A-Ans2b. [It becomes an absolute nuisance. You end up with things like the joiners first-fixing before the windows are fitted, which means the plot's not water-tight; the plumbers and sparks in together and pinching each others pre-drilled service holes in the joists; and the plasterers boarding the upstairs ceilings and one side of the stud walls while the building inspector's walking around the plot doing his pre-plaster inspection, which isn't ideal really. Then you might double up on your plasterers to get it plastered quicker but still have to put joiners second-fixing downstairs while they're finishing off skimming upstairs while the plumber's fitting his heating and sanitary wherever he can—it becomes madness at times, sheer madness.] **T2**

A-Ans2c. The implications for quality are the worst thing for me.

A-Ans2c. [I can deal with the long hours and trades having a 'barny' with one another but it's the bad effect on quality that really ****** me off at those times.] **COMP1/QUAL2**

A-Ans2c. [A lot of the time I do think these situations occur simply because the construction director won't say no to the MD. I wish he'd just once say 'no, you can't have that plot in three weeks, it's impossible. You'll have to have it next month instead or it'll be a load of ****' or something like that.] **PROP2**

A-Ans2d [The subbies themselves all basically do as they're told.] **QUAL3**

The lads on site may moan about working weekends but they do get a bonus of about £25 per man, per day, for coming in, on top of whatever price-work they do so at the end of the day they are earning more money for a few weeks. Ehm,

A-Ans2d [as for their bosses, well, they moan too but at the end of the day they daren't refuse because they're **** scared of not getting any more work off us. I don't think any of them are bothered about quality really because if it is bad they can just blame us for making them rush it.] **QUAL3**

A-Ans2d [I'd have to say at times there are problems with health and safety too. Ehm, it's unavoidable really if you've got men working outside normal working hours. There's just bound to be times when someone ends up working on site without the benefit of a first aider present or even access to the first aid kit or toilets or washing facilities if the cabins are all locked up. Your painter can often end up as a lone worker too because they'll just lock themselves in a plot and stay to all hours to get it finished.] **T3**

I suppose it's quite dangerous practice when you stop and think about it.

A-Ans2e. [Well none of them use supervisors or foremen or anything like that, not really, no.] **QUAL2**

I mean, if I have any issues with any of the lads on site regarding quality I'll ring their bosses and get them out to look at it with me and they always 'play ball', to be fair. Anyway, all you have to do is threaten not to pay 'em until they do come out and they'll soon turn up to sort it.

A-Ans2f. [Well according to our company policy we are supposed to do them once a month and we have a pro forma on which to record the meeting and report it back to head office but no manager does them on a regular basis as far as I'm aware.] **COMP3**

I will always do one at the beginning of a new site as a way of introducing myself face-to-face to the subcontractors who I don't know or haven't used before. I find this initial meeting useful as it puts a face to a name and makes it easier in the future when you're dealing with them over the phone.

A-Ans2f. [Doing them each and every month though just isn't practical especially at crucial times of the year when you're mad busy because you just don't have the time and neither have they really.] **COMP1/COMP4**

I mean you're talking to all the subbies on a daily basis anyway because the job's moving so fast so most of the managers don't see the point of them really and the contracts manager and construction director only seem to mention them when things are quiet and they have time to 'nit pick' over things—just like they do with the health and safety policy.

A-Ans3. Good, very good.

A-Ans3. [We have a good understanding. I know what they expect and I've got high expectations, as have they, so, ehm, as long as we work together, we do a good job.] **PROP2**

A-Ans3a. [I wouldn't say the relationship ever becomes tense or bitter as such, ehm, not really.] **PROP2**

I mean it can be difficult when you've got a lot of plots, a lot of completions in one particular month, ehm, but as long as we all work together and pull together, the help is already there so all you've got to do is ask for it.

A-Ans3b. I think that naturally there is some,

A-Ans3b. [there would be a small amount of conflict there wouldn't there? Ehm, with them pushing to get profit by achieving the sales completions, which obviously increases the chances of them getting full bonuses. I wouldn't say it was unhealthy though, to have that.] **PROP2/PROF3**

A-Ans3c. Well

A-Ans3c. [my bonus is based on my NHBC items, reportable items that you get when the building inspector records defects he's spotted so it can affect how much bonus I get if they force me to hand over a plot with loads of items on it instead of giving me extra time to get them sorted.] **PROF1/T1**

A-Ans4. [Overall I would have to say we are given enough time in the build programme to build a good house because quality seems to come before productivity at this company.] **T1**

Occasionally you are given less time than others depending on the plot and upon how and when it's been sold during a particular month and when the completion needs to be in for.

A-Ans4. [Time pressures can be more severe on one plot than it could be for another but, you know, they're nearly always achievable.] **T1**

A-Ans4. [It's often last minute sales and stuff like that the cause the problems with time constraints.] **T4**

If we sell a plot and it has to be in within 4 to 5 weeks and it's achievable, we will certainly do our best to achieve it.

A-Ans4. [They don't usually give you impossible scenarios. We can usually achieve what were asked to achieve.] **PROP2**

I mean

A-Ans4. [I have had situations in the past where I've been extremely rushed and where, say, a 16-week build, we've had to try and achieve it in 12 weeks. Well naturally, you know, I've not been happy with that, with having to do that, because with them extra 3 or 4 weeks, you can get things more pristine.] **T1**

A-Ans4. [The funny thing is though, when they do let you stick to the programme the pressures are probably greater because the expectations are higher because you've been given more time.] **PROP1**

Motivation is a key in this. You need to be motivated and ensure that you stay motivated, even though you might have a bit more time to complete it.

A-Ans4. [You see, the overall expectation is far greater because you've had more time and once you've handed over the plot complete, they expect it to be to a really good standard, with greater emphasis on quality.] **PROP1/T1**

A-Ans4a. [Probably the biggest negative aspect of working here, compared to other house builders I've worked for, is that cost controls are extremely tight.] **COMP2**

Everything you order is scrutinised and anything extra you ask for is questioned; even something like an additional broom for sweeping the welfare cabins out with.

A-Ans4a. [It becomes really frustrating and annoying at times, the way they 'penny-pinch', especially if they're asking you to build a plot in 30% less time than normal but they don't want to pay for temporary lighting even though we haven't got electric to the plot yet.] **COMP2**

They expect miracles on a shoestring budget but we usually get there in the end, which of course means that the situation never changes.

A-Ans4a. [The more often you do it that way, the more often they want it that way.] **PROP2**

A-Ans5. I haven't really experienced that within this company, no. With other companies I've worked for, yeah, but not here really, not with the current structure we've got.

A-Ans5. [I think my finished units are always very good. We do have various stages and various viewings by both the contracts manager and construction director. We all look at it, you know, at various stages of the actual finishing stages and by the time we come to actually handing it over, I would say my standard's quite good.] **QUAL2**

A-Ans5. [You do get times when you want another day or two to finish a plot off but you're told no because they need it by that date to get the money in on time for the year-end deadline] **PROF1**

and

A-Ans5. [sometimes it is difficult but the way I look at it, if it has to be in for that month, it has to be in. If it means putting a few extra hours in to achieve that, well we have to do it.] **COMP1**

If, you know, nine times out of 10 it can be done, then yes, we certainly try our best to achieve it, although sometimes it's extremely difficult.

A-Ans5a. [You do sort of feel compelled to do it, basically because it's the worry about losing your job and we've had that in the recent past.] **PROP1**

The demand can be so high that, ehm, that we would be rushed because of demand.

A-Ans5a [Besides that, if we didn't co-operate and we didn't cope with demand, then we would either not get our bonus or our bonus would be extremely reduced.] **PROF3**

A-Ans6. [It's down to the site manager mainly. The contracts manager will have a bob round about once a week when he comes to site but apart from that it's down to me. I do all the inspections really. We don't have a quality control manager or anything like that.] **QUAL2**

A-Ans6a. [I don't believe I ever miss any of the required inspections out.] **QUAL2**

Like I mentioned earlier, our bonuses, the site managers' bonus that is, are only based on the amount of reportable items you get so

A-Ans6a. [It's in your own interest to thoroughly check plots before key-stage inspections by the NHBC inspector, otherwise you may end up with no bonus.] **PROF3**

A-Ans6b. [During busy periods like at year-end when you've got some tight plots because they've been thrown on you at the last minute, you might not get round all the plots as often as you'd like or spend as much time in each one as you normally would but you still make sure you do your checks, even if it means you being on site at six o'clock in the morning.] **PROP3/QUAL2**

Having said that,

A-Ans6b. [I've had many occasions where the NHBC inspector has turned up to do a CML while I'm half way through doing my final checks. You just have to keep your fingers crossed that you haven't overlooked any major defects when that happens.] **QUAL2**

A-Ans6c. [It really depends on which inspector you get as to how thorough he is. Some can be real pernickety ******** and others quite lax. My inspector on here is a bit of both; he can be okay one day and a complete 'jobsworth' the next. At times like year-end it also depends on how many visits they have on. With a lot of builders having they're year-end around the same time, if they're all wanting CMLs on the same day then he obviously won't have as much time to spend in each one] **QUAL2**

so that can work to your advantage.

A-Ans6d. [We have a procedure set out in the quality management books.] **QUAL2**

A-Ans6d. [What's supposed to happen is that a few days before the CML is done, the site manager and the sales negotiator 'snag' the plot and the site manager sorts the snags before the CML.] **QUAL1**

A-Ans6d. [After the CML is done, the sales people make an appointment with the client to come and view the property in what we call a 'home demo' but in the book it's called a 'Pre-Handover Viewing' and if the client picks up any other snags, the site manager is supposed to try his best to get them done before the client moves in.] **QUAL1/PROF2**

A-Ans6d. [At the actual handover of the property to the client, which the book calls a 'Key Release', they're again asked to note down any defects but also to sign off where issues have been resolved.] **QUAL1**

Then

A-Ans6d. [seven days after they have physically moved in, the site manager does another inspection with the client and once they've signed off the book to say he's done them the book goes into the office and the customers then have to contact head office direct with any problems.] **QUAL1/PROF2**

I say

A-Ans6e. [that's what's supposed to happen because it very rarely pans out that way. Firstly, the sales people – I won't say sales women because we do have a man as a sales negotiator – the sales people are all 'bone idle', especially when it comes to filling out the 'Quality Management' books. You see, it's their responsibility to do them. They have the books from day one and it stays with them until I do my seven-day call. I can't remember the last time I did a pre-CML check with a sales negotiator. I do, do them of course but I do them on my own.] **T4**

The other thing is that,

A-Ans6e. [especially at year-end, you often have the scenario where clients are moving in the same day you get the CML or the house was only actually completed on the day of the CML so you obviously can't follow the procedures in the book.] **PROP3**

A-Ans6e. [It's all a good idea in principle but I think it's mainly just a token gesture to try and make the company look efficient.] **QUAL2**

A-Ans6f. [Some defects often have to be rectified after completion but it's nearly always due to time limitations. I don't think it's ever about the site manager just not trying hard enough.] **QUAL1/T1**

The other thing is that all house builders work within tolerances so there are certain things we don't have to attend to but the clients often think we do. If it's something that's only half-an-hour of a job, such as a painter touching up after they've had carpets fitted, then you'll often just do it as sort of a gesture of good will to try and get the customer on your side and establish a good relationship with them from the start.

[I would say we probably do around 30% of the snags after completion.] **QUAL1**

A-Ans7. I think that house buyers are more demanding of good quality than in the past, yes, very much so, especially with the Internet and programmes on TV and in the general media, its ehm, it is becoming more difficult as time goes by. I mean, the media in general over the past few years has promoted a bad image and people these days are expecting far more for what they're actually paying out, whereas in the past they were just, you know, happy to purchase a plot for instance and just go along with it but nowadays, they just, they want the best for what they're paying for.

A-Ans7a. [I think the quality of the actual build has remained the same, if not improved; it's the expectations of the actual client, the purchaser that has risen. Their expectations have got far greater because they want value for money and they want what they want and they seem to know a lot more about what they want through the use of the internet and the media.] **PROF2**

Ehm, I think our attitudes have remained the same and if the job, if the build is actually built to the way it should be along the process of the NHBC inspector looking at it at the various key stages and we all know as site managers what the company expects, then I'd have to say that quality has probably improved.

A-Ans7b. [If the defect's there to be reported, then the customer reports it, so yes, I suppose they do tend to report more defects now than a few years ago.] **QUAL1**

A-Ans7c. [The home demo is usually the first chance the customers get to see their property. It's the best time really; to see it when it's completely finished because it gives them a better impression of what they've actually bought. You sometimes get the odd one wanting to view the plot at first-fix or when it's plastered but site managers discourage this because clients end up walking round the house pointing at all sorts saying 'this isn't finished' or 'that isn't finished' and you feel like saying 'I know it's not finished you ****** fool, we're still building the ******* thing'.] **PROF2**

A-Ans8. Ehm, not really sure about that one.

A-Ans8. [We sometimes get clients giving us bottles of wine and stuff as a thank you if they're really happy with their new home and some do send letters in to head office praising the site staff] **PROF2**

but apart from that I don't really know. We've actually had relatives of people who work in our head office buy houses off us but that's because they think they'll get a good deal more than anything.

A-Ans 8a. [You only really get to hear from the ones who threaten to go to the press or threaten to tell everyone they know not to buy a ************ house but I don't think that's changed much over the years.] **PROF2**

A-Ans 8b. Ehm,

A-Ans 8b. [they used to reckon one in every 10 home buyers would be a 'customer from hell' so judging from that I don't think we get that many really bad ones but you never know what people say to their family and friends, do you?] **PROF2**

A-Ans9. [I was nominated this year but I haven't heard anything of that yet and as far as I know at least two other site managers have been nominated for 'Pride in the Job' awards.] **QUAL2**

A-Ans9a. [For some companies I definitely know they are run differently but at this company, no, and that's mainly due to the cost constraints. We don't have a great deal of money to spend on the likes of fancy sales and site compound areas and other presentation ******** like other companies do.] **COMP1**

We do still try to achieve the standard of quality that would enable us to win an award but it's not really possible to achieve the overall standard that the NHBC is looking for when it comes to awards without spending loads of money 'tarting' the site up. Besides,

A-Ans9a. [everyone in the game knows that the biggest house builders take the **** bosses on golfing weekends and trips abroad and what have you so a lot of it's plain ******* really and little to do with the way they actually build the majority of their houses.] **QUAL2**

A-Ans10. Yes, definitely. I would say all of them. I know all the managers here personally and I can't think of any one of them, off the top of my head, who would disagree with anything I've said.

*File 9: a new file, comprising tables*

| Data coding number | Main category headings | Sub-category headings | Frequency counts | Literature sources | Observations, implications or interpretations | Data consistencies | Data inconsistencies |
|---|---|---|---|---|---|---|---|
| 1 | Propensity | Culture | 155 | Atkinson (2002, 1999), Tam *et al.* (2000), McCabe *et al.* (1998), Reason (1998) | If a propensity exists among the site managers then it would seem to stem from higher management targets. Various forms of intimidation are used to pressure the site managers. | Excessive pressure is placed on site managers to achieve company targets. Self-esteem is a prime motivator for site managers. | Job security is not the overriding factor regarding propensity. Some pressure is self-inflicted. |
| 2 | | Hierarchy | 88 | Liu (2003), Reason (1995), Morris (1994) | Senior managers and directors operate from a position of self-interest motivated by personal monetary gain via bonus payments. A lack of respect for superiors prevails among site managers. | Most managers experience strained relations with senior managers during busy periods. | A few managers claimed their relationships with hierarchy is unaffected by pressure. |

(*continued*)

(*Continued*)

| Data coding number | Main category headings | Sub-category headings | Frequency counts | Literature sources | Observations, implications or interpretations | Data consistencies | Data inconsistencies |
|---|---|---|---|---|---|---|---|
| 3 | | Specific periods | 105 | | The PHS as a whole, experiences unusually busy periods in an annual and/or bi-annual cycle. There is a lack of support from higher management during busier times. | Adverse working practices are frequently adopted during such times. | Some become institutionalised by the casual attitudes of hierarchy towards product quality and H&S. |
| 4 | Completions | Workload | 117 | | Excessive workloads and unsociable hours are often imposed on site managers during year-end and half-year-end periods. This has detrimental psychological effects on many site managers on a bi-annual cycle. | Unfair expectations are placed upon site managers during crucial periods. | A few claimed to be comfortable with their workload during busy periods. |
| 5 | | Resources | 16 | Langford *et al.* (2000) | Additional resources are available when needed. | Extra resources cannot compensate for lack of time. Control of resources is difficult for site managers. | Some site managers feel that being given extra resources increases the pressure to achieve unrealistic targets. |

| | | Count | References | | | |
|---|---|---|---|---|---|---|
| 6 | Supply chain | 44 | Barker & Naim (2008), Housing Forum (2001), Naim & Barlow (2003) | The fragmented nature of the PHS is reflected in a dysfunctional supply chain. Lead times of manufacturers are inconsistent. | Problems with delayed or incorrect materials deliveries are rife. | Utility providers pose the most problematic external danger for house builders. |
| 7 | Communication | 13 | Ozaki (2003) | There is an inherent lack of formal communication between site managers and subcontractors. | Formal site meetings with subcontractors are not frequently conducted. | There are failures in upwards communication flows between site managers and superiors. |
| 8 | Quality — Defects | 142 | Kim et al. (2008), Sommerville (2007), Garrand (2001), Kletz (2001), Reason (2000, 1990), Minato (2003), Roy & Chocrane (1999), Atkinson (1998) | Increased defects are mainly the result of compressed construction duration and inadequate inspection times. Tradespeople can use reduced timescales as an excuse for shoddy work. | The site managers all displayed a conscientious attitude towards product quality. | Houses can always be built to a reasonable standard. |
| 9 | Quality control | 81 | Barker (2004), Cheng et al. (2002), Tam et al. (2000) | Third party **** building inspections have limited significance for product quality. Doubt exists over the validity of **** quality **** awards. | There is an absence of any effective, standardised system of quality control among private house builders. | Even where formal quality control procedures exist they are usually compromised by completion rates. |

*(continued)*

(Continued)

| Data coding number | Main category headings | Sub-category headings | Frequency counts | Literature sources | Observations, implications or interpretations | Data consistencies | Data inconsistencies |
|---|---|---|---|---|---|---|---|
| | | | | | Very few subcontractors employ supervisors to ensure satisfactory quality is achieved. | | |
| 10 | | Workforce | 213 | Power (2000), Tam et al. (2000) | Subcontractors are more concerned with securing future tenders than with producing quality work. Unreasonable working hours and timescales imposed from above, lowers moral and causes friction between site managers and tradespeople. | The use of subcontracted labour is prevalent within the PHS. | A preference for directly employed tradespeople exists among site managers; even with those who had not used this type of labour for some time. |
| 11 | Profit | Budget targets | 72 | Rosenfeld (2009) | There is an unstated assumption that profit must ultimately take priority over quality. | Senior managers and directors have a vested interest in securing overall budget forecasts for the company. | Site managers do not have a vested interest in securing overall budget forecasts for the company. |

| No. | Category | References | | | |
|---|---|---|---|---|---|
| 12 | Repeat business | 4 | Mbachu & Nkado (2006), Power (2000), Zeithaml et al. (1996), Hoyer & MacInnis (2005), Pitcher (1999) | The negative perception of new build housing portrayed by the media is unfair. Site managers have only a vague awareness of customer satisfaction levels. | Most home buyers have unrealistic expectations. | The link between customer satisfaction and customer loyalty is given little or no credence within the PHS. |
| 13 | Bonus payments | 69 | | The bonuses of senior managers and directors are mainly profit-related. The different criteria by which most site managers' bonus payments are evaluated causes inevitable concern | The bonuses of site managers are often dependent on not receiving reportable items (recorded defects) from the building inspector. | Site managers with bonuses scored according to similar criteria as senior managers experience less conflicts of interest. |
| 14 | Build programmes | Time | 20 | Condensed build programmes is the main factor in the inability to produce good quality housing. | No allowances are made for delays. | The build programmes are adequate for most of the year. |
| 15 | Working conditions | 22 | Roy et al. (2003), Morris (1994), | During year-end and half-year-end periods, the working practices adopted by most private housing companies is not conducive to good quality. | Overcrowding and out-of-sequence works is commonplace during busy periods. | When houses are built to a unreasonable programme, the expectations of good quality can then become unrealistic. |

*(continued)*

*(Continued)*

| Data coding number | Main category headings | Sub-category headings | Frequency counts | Literature sources | Observations, implications or interpretations | Data consistencies | Data inconsistencies |
|---|---|---|---|---|---|---|---|
| 16 | | Health & safety | 22 | Reason (1997), Whittington *et al.* (1992) | Unsupervised out-of-hours working is often unavoidable but is a major health and safety concern among site managers. | Overcrowding and reduced supervision are the main causes of increased risk. | Effective forward planning combats health and safety concerns. |
| 17 | | Sales issues | 53 | CIH (2008), Forsythe (2007), Livette (2006) | There is a philosophy among the hierarchy to complete the sales and worry about quality once the money is in the bank. | Last-minute sales are one of the biggest causes of rushed completions. | Sales staff are considered incompetent and contribute to delays in production during the busiest times. |

*File 10: the narrative*

It is suggested that a culture of putting completions before quality exists, and the data set illustrates that it originates with top-level management and passes down through the chain of command to site level. There is a perception that directors are aggressive in pushing for completions, and they remain silent about the consequences for quality, but realistically know it will suffer. The systems are such that if problems arise after handover, site managers (SM) can be blamed. Contract managers, in their position between directors and SMs, are reluctant to say no to directors, since this may prejudice their positions. Directors seem motivated by 'brownie points' and bonus schemes that are based upon payment for completions. Directors' remuneration packages may be substantially based upon achieving targets for completions. There appears to be a common belief among SMs that the completion targets set by directors are motivated by their self-interest. Tam *et al.* (2000) lend weight to these grievances by positing that culture-related issues are the most important factors affecting construction quality.

Most SMs expressed a resistance to pressure placed upon their job security, at least in principle, but as one SM put it 'there's resistance there from the site managers but you have to do as you're told at the end of the day and get them through, so you just go for it'. Half of the managers voiced concerns over job security and most of them also hinted at issues of self-esteem as part of working within a team. One SM expressed a wish to be well thought of by superiors, and reflected this is just part of human nature.

Adverse working practices are clearly commonplace at specific times of the accounting year, which are generically termed 'year-end' and 'half-year-end', when bonuses are calculated. As one SM commented 'it's just the completion periods that cause the problems really. I'd say on average there's a period of about a month really, twice a year, prior to year-end and half-year-end periods, when things get crazy'. One SM conceded that most of the time SMs are fighting losing battles if they try to delay completions, and they may be tempted to push completions through, and if there are problems later, blame them on directors and contracts managers; however, this can be quite demoralising. Interestingly, one manager believed some of the pressure was self-inflicted, stating that SMs should do more checking of quality and should not use time as an excuse for not doing so; SMs should work hard all the time and not 'slack-off' as is sometimes the case. Pressure to achieve completions gets passed on to tradespeople. Subcontractors are asked to start early, work late or all day Saturday and all day Sunday; not just a 'couple of hours' on a Saturday morning. Many trades may be working in one property at the same time, and this has consequences for productivity, quality and safety.

Rather alarmingly, a few of the interviewees considered their contracts managers to be ineffective at their jobs, as demonstrated through comments such as, 'my contracts manager isn't particularly good under pressure so he tends to stay away. He's pressurised on all the sites, everybody's chasing him'. Directors could also be said to be somewhat aloof from the realities of trying to build houses under severe time constraints and to an acceptable standard of quality. Some SMs think that contracts managers do not appreciate site problems or turn a blind eye to them. Even though they have site experience, they get 'removed' from reality after a few years. One SM reported that when things are tough contracts managers can 'sit back like the generals on the hill watching everybody else charge around and they won't even answer their 'phones half the time' and when SMs ask for help the response is 'get it sorted, everybody else is in the same boat, just shut up and get on with it'. Only one SM claimed to be able to successfully resist condensed build programmes and gave an example of telling head office that completions would not be achieved on Friday, but it will be on the following Tuesday or Wednesday. Only one manager claimed not to have any relationship problems with senior managers, 'I tend to be strong enough to get my own way with my contracts manager and even the director'. However, for most of those questioned, successful resistance was rare.

Bonus systems for some SMs may be set with two competing interests. One element of a bonus may be based on having few defects on snag lists produced by building inspectors, whilst the other element is based on getting the completion itself. A bonus can comprise a significant proportion of SMs' remuneration packages; therefore, they can be as highly motivated as directors and contracts managers to achieve completions. Despite the obvious lure that significant amounts of money can present, the general impression from the data is that the main 'bone of contention' for SMs is the inability to build houses to a satisfactory standard of quality.

Pressure applied by directors appears to have disturbing psychological consequences for some SMs, such as negative impacts on working relationships as well as their personal and family lives. One SM reported 'you start to lose the team... and seem to lash out at them because you're frustrated, with the office, with the senior managers and director'. Another SM recalled 'the amount they ask you for at year-end, you can't physically do it; there's never enough hours in the day. . . . I won't be taking any breaks during the day and I'll have days when I won't have a lunch and . . . I'll have some weeks where I'll turn in at half six in the morning, before everyone's got there, . . . and then I'm going home and waking up in the middle of the night thinking about plots . . . come the afternoon, sometimes your brain goes into 'meltdown' . . . then when I go home the missus is telling me I look gaunt and losing too much weight because you're just on the go so much all the time and you do start making mistakes yourself'.

This apparent transfer of disregard for product quality is perhaps reflected by Liu (2003, p. 149) who details how people become institutionalised through hierarchical power arrangements that affect their behaviour by guiding their thoughts and beliefs to create group values. Hence, it could be said that the attitudes and behaviours of senior managers and directors within the private house building sector towards quality will be mirrored among tradespeople (and some SMs) with resultant negative effects on quality.

Opinions vary concerning the additional workload imposed on managers when timescales are reduced. One SM claimed that although fast-track build completions were attainable, they did require the manager's full attention because 'if you're doing a house in two weeks from plaster to completion, you have to sit on that house and that's all that's on your mind and you kind of push everything else to one side a little bit to give that house the 100% attention it needs'. This suggests that while it may be possible to complete houses in this manner and to an adequate standard of quality, it could only be done on a limited number of plots before quality standards begin to be compromised.

Despite most of the managers indicating that additional resources such as assistant managers and labourers were readily available during busy periods if needed, none of them felt that this compensated for the lack of time in the build programme. In cases where there is initially adequate time, there can be problems with resources if they are taken off site to attend to urgent completions elsewhere. Trying to achieve completions can be frustrated by problems with material deliveries. If customers are allowed to pick their own kitchens, the delivery period may extend beyond the proposed completion date. Materials may be late or be delivered short or damaged, and it can be difficult trying to 'rob' items from other plots. Another widespread problem is the delay with service connections (gas, water and electricity) to properties, which can impact on production and quality in a variety of ways.

There appears to be an inherent lack of communication between SMs and subcontractors. Only three of the eight SMs held regular formal site meetings with subcontractors; they did claim the meetings to be worthwhile. The majority seem to view them as an unnecessary chore with little or no benefit, and argue that neither they nor the subcontractors have time. It would appear that most SMs prefer a more *ad hoc* approach to communication. It may be argued that the general fast pace of the private housing sector does not lend itself to such organisational 'tools' as it does with the slower, more controlled process of the 'construction' industry at large.

There is an appreciation that whilst money is the key driver, maximum profits are primarily derived from positive cash flow and not efficiency or lowest costs on site. This is particularly the case if high value properties are being completed in clusters, e.g. four properties at £0.25M each, total income £1M. This sum of money in the bank means far more than relatively minor problems with quality. It is perceived by some to be the lesser of two evils to have hassle with customers and work inefficiently to rectify defects after customers have moved into a house, rather than alternatively, not to have the cash. It is probably on this basis that owners of companies reward directors based upon housing completions rather than on quality.

# Appendix E: using Excel for charts, descriptive tests and inferential tests

Spreadsheet software is virtually essential for speed, and to enable you to present the data appropriately. Using copy and paste functions, charts are easily transported into Word documents. A summary of data sets can be presented on a spreadsheet, and included in the main body or an appendix. In the spreadsheet cells, it is usual to code qualitative labels with numbers, e.g. male = 1, female = 2.

## Four charts

Instructions are given to plot four charts that are often used: frequency histograms (columns), line diagrams, pie charts and scatter diagrams. For all charts, use the chart wizard function. The default charts presented can all be adapted to suit personal preference for patterns, fonts, and properties, using the 'format chart' function.

### Frequency histograms

Table E.1 illustrates how the data should be presented in the cells in the spreadsheet. The data are based on Table 7.5 columns 3 and 4, and the frequency histogram itself is shown in Figure 7.1a.

Before entering chart wizard, start with data cells in the range A1 to B10 all highlighted. If the relevant cells are not adjacent on the spreadsheet, keep the control key pressed down while highlighting with the mouse. Click on chart wizard > column > next > columns (not rows) > next > titles > type in relevant details for chart title, category (X) axis, and category (Y) axis > legend > un-tick show legend > finish.

**Table E.1**  Typical data for plotting a frequency histogram.

|  | A percentage scores | B frequency counts for each class interval |
|---|---|---|
| 1 | 30–34 | 1 |
| 2 | 35–39 | 1 |
| 3 | 40–44 | 2 |
| 4 | 45–49 | 5 |
| 5 | 50–54 | 8 |
| 6 | 55–59 | 6 |
| 7 | 60–64 | 3 |
| 8 | 65–69 | 2 |
| 9 | 70–74 | 1 |
| 10 | 75–79 | 1 |
|  |  | 30 |

### Line diagrams

The following instructions are based on the original data set in Table 7.8, represented in Table E.2. The line diagram is shown in Figure 7.5. Before entering chart wizard, start with

**Table E.2** Data set adapted from Table 7.8. Tender and cost price indices (BCIS, 2010).

|    | A | B | C |
|----|------|--------------------|-------------------|
| 1  | Year | Tender price indice | Cost price indice |
| 2  | 1999 | 151 | 183 |
| 3  | 2000 | 161 | 191 |
| 4  | 2001 | 174 | 196 |
| 5  | 2002 | 187 | 205 |
| 6  | 2003 | 197 | 215 |
| 7  | 2004 | 213 | 227 |
| 8  | 2005 | 224 | 241 |
| 9  | 2006 | 230 | 255 |
| 10 | 2007 | 245 | 267 |
| 11 | 2008 | 245 | 282 |

data cells in the range B2 to C11 all highlighted. Click on chart wizard > line > next > in the data range tag, rows > in the series tag note in the series box the titles series 1 and series 2 > in the 'name' box type tender price indice > add > in the series box highlight series 3 and remove > in the 'name' box type cost price indice > add > in the series box highlight series 2 and remove > place the cursor in the cell 'category (X) axis labels' > highlight cells A2 to A11 > next > titles tag, type in chart title = BCIS tender price indice and cost price indice > in the category (X) axis cell type year > in the value (Y) cell type indice, 1985 = 100 > next > finish. To adjust the vertical scale on the finished line diagram, double click on the numbers on the vertical scale > scale tag > minimum = 100.

### Pie charts

Table E.3 presents new hypothetical data for age of respondents, in three categories, perhaps from a survey. It is shown as it should be presented, in the cells in the spreadsheet.

Before entering chart wizard, start with the six data cells in the range A1 to C2 all highlighted. Click on chart wizard > pie > in the data tag, tick 'rows' > go to the series tag > to add a title, in the 'name' box type 'age of respondents' > add > remove the 'series 1' label > next > in the titles box 'age of respondents' will appear > in the data labels tag, tick 'category name', 'value', and 'percentage' (do not tick series name) > in the legend tag, ensure 'show legend' not ticked > next > as object in > finish. Figure E1 illustrates the pie chart.

### Scatter diagrams

Table E.4 illustrates how the data should be presented in the cells in the spreadsheet. The data are based on Table 8.26, and the scatter diagram itself is shown in Figure 8.4.

**Table E.3** Data in Excel to produce a pie chart.

|   | A | B | C |
|---|-----------|---------------|------------|
| 1 | < 35 years | 35 to 54 years | ≥ 55 years |
| 2 | 8 | 14 | 10 |

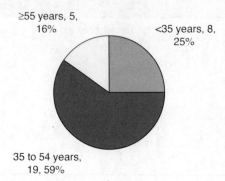

**Figure E.1** Pie chart after following Excel instructions.

**Table E.4** Illustration of data in a table to produce a scatter diagram in Excel.

| | A pre-start percentage change in cost predictability | B ditto in time |
|---|---|---|
| 1 | + 5.0 | + 1.0 |
| 2 | 0 | + 5.0 |
| 3 | 0 | 0 |
| 4 | −1.1 | + 1.1 |
| 5 | + 15.3 | + 10.5 |
| 6 | + 2.5 | 0 |
| 7 | + 1.0 | 0 |
| 8 | + 8.3 | + 12.5 |
| 9 | + 6.2 | + 3.0 |
| 10 | + 4.8 | + 12.0 |
| 11 | −3.0 | 0 |
| 12 | + 4.5 | + 8.5 |
| 13 | −2.8 | 0 |
| 14 | −7.2 | −2.0 |
| 15 | −1.0 | 0 |
| 16 | 0 | −3.3 |
| 17 | −0.1 | + 2.1 |
| 18 | 0 | + 5.0 |
| 19 | + 6.6 | −1.0 |
| 20 | −0.2 | 0 |
| 21 | −2.2 | + 5.0 |
| 22 | −5.2 | 0 |
| 23 | + 8.2 | + 15.0 |
| 24 | + 2.0 | + 8.0 |
| 25 | + 1.2 | + 2.0 |
| 26 | + 3.2 | + 5.0 |
| 27 | + 10.0 | + 20.0 |
| 28 | −3.0 | + 2.0 |
| 29 | −2.2 | + 3.0 |
| 30 | −2.9 | 0 |

Start with data cells in the range A1 to B30 all highlighted. Click on chart wizard > scatter > next > columns (not rows) > next > titles > type in relevant details for chart title, category (X) axis 'cost predictability', and category (Y) axis 'time predictability' > legend > un-tick show legend > finish.

## Eleven descriptive tests

The 11 tests described are: count, sum, mean, median, mode, standard deviation, variance, maximum, minimum, range, and confidence interval. With the Excel data sheet open and data entered, the first instructions for all eleven tests are:

in the main dialogue boxes labelled home, insert, page layout, formulas, data, review, view > go to formulas

To illustrate the use of the tests, the data set in Table E.5 is used; it is based on Table 7.4. Before clicking on the 'go to formulas' dialogue box, place the cursor in the cell on the spreadsheet, where the answer is required, e.g. in Table E.5, for the count that will be in cell B31, and for sum in cell B32

The five descriptive tests (count, sum, mean, maximum and minimum), are accessible thus:

Click on the 'go to formulas' dialogue box. Click on the test required. Excel automatically identifies a range of cells; ensure it is B1 to B30, and press 'enter'. The result is given in the appropriate cell, e.g. count = 30, sum 1628, mean 54.27, median 53, maximum 75, minimum 34

For the range statistic, take the difference between the maximum and the minimum, using the Auto-sum function, ensuring that the formula reads '=SUM(B38–B39)'

For the remaining five descriptive tests (median, mode, standard deviation, variance, and confidence interval), access is made thus:

Click on the 'go to formulas' dialogue box > insert functions > statistical > scroll down to find the test required > (note standard deviation is abbreviated to STDEV, variance to VAR, and confidence interval to CONFIDENCE) > OK > place the cursor in the function arguments dialogue box, where labelled 'number 1' > using the mouse go to the spreadsheet main page and block in the cells B1–B30 (this will give a reading of B1: B30) > the result will appear.

For each of the tests, in the 'insert function' and 'function arguments' dialogue boxes respectively, the following text automatically appears:

MEDIAN: returns the median, or the number in the middle of a set of numbers.
    Number 1, number 2, . . . are 1 to 30 numbers or names, array or references that contain numbers for which you want the median

MODE: returns the most frequently occurring, or repetitive, value in an array or range of data
    Number 1, number 2, . . . are 1 to 30 numbers or names, array or references that contain numbers for which you want the mode

STDEV: Estimates the standard deviation based on a sample

Number 1, number 2, . . . are 1 to 30 numbers corresponding to a sample of a population and can be numbers or references that contain numbers

VAR: Estimates the variance based on a sample

1 to 30 numeric arguments corresponding to a sample of a population

CONFIDENCE: Returns the confidence interval for a population mean

**Table E.5** Data set based on Table 7.4 to illustrate Excel calculation of 11 descriptive tests.

| | A | B |
|---|---|---|
| 1 | R1 (respondent 1) | 40 |
| 2 | R2 | 44 |
| 3 | R3 | 53 |
| 4 | R4 | 47 |
| 5 | R5 | 63 |
| 6 | R6 | 56 |
| 7 | R7 | 56 |
| 8 | R8 | 34 |
| 9 | R9 | 50 |
| 10 | R10 | 47 |
| 11 | R11 | 53 |
| 12 | R12 | 59 |
| 13 | R13 | 59 |
| 14 | R14 | 63 |
| 15 | R15 | 63 |
| 16 | R16 | 47 |
| 17 | R17 | 69 |
| 18 | R18 | 47 |
| 19 | R19 | 38 |
| 20 | R20 | 53 |
| 21 | R21 | 69 |
| 22 | R22 | 72 |
| 23 | R23 | 75 |
| 24 | R24 | 59 |
| 25 | R25 | 56 |
| 26 | R26 | 53 |
| 27 | R27 | 53 |
| 28 | R28 | 50 |
| 29 | R29 | 47 |
| 30 | R30 | 53 |
| 31 | Count | 30 |
| 32 | Sum | 1628 |
| 33 | Mean | 54.27 |
| 34 | Median | 53 |
| 35 | Mode | 53 |
| 36 | Standard deviation | 9.79 |
| 37 | Variance | 95.79 |
| 38 | Maximum | 75 |
| 39 | Minimum | 34 |
| 40 | Range | 42 |
| 41 | Confidence interval | 3.65 |

Alpha is the significance level used to compute the confidence level, a number greater than 0 and less than 1. Standard Dev is the population standard deviation for the data range and is assumed to be known. Size is the sample size

In the case that you wish to perform the same test on different columns of data, use the copy and paste functions; there is no need to go through the routine of 'go to formulas > insert function > function arguments . . .'. When using copy and paste, you need to make sure that the correct array of cells are identified.

## The five inferential tests

### Pearson's chi-square test

Table E.6 is used to illustrate some data based on Tables 8.3 and 8.4. In the main dialogue boxes labelled home, insert, page layout, formulas, data, review, view > go to formulas > more functions > statistical > CHITEST > the dialogue box 'function arguments' appears with the text 'returns the test for independence: the value from the chi-square distribution for the statistic and the appropriate degrees of freedom' > place the cursor in the cell labelled 'actual range' and the following text appears 'actual range is the range of data that contains observations to test against expected values' > using the mouse go to the spreadsheet main page and block in the cells A1, B1, A2 and B2; this will give a reading in the actual range cell of A1: B2 > place the cursor in the cell labelled 'expected range' and the following text appears 'expected range is the range of data that contains the ratio of the product of new totals and column totals to the grand total > block in the cells C1, D1, C2 and D2; this will give a reading in the expected cell of C1: D2 > formula result will appear, with the $p$ value given thus: formula result $p = 0.023$. Note that Excel does not use Yates' correction; by manual calculation in Table 7.13, the chi-square value is calculated to be 2.98 and not significant with $p \leq 0.05$. Step 3 in the calculation, without Yates' correction, gives a chi-square value of 4.36, which is significant with $p \leq 0.05$, and a precise value by Excel of $p = 0.023$.

### Wilcoxon test, unrelated t-test and Mann–Whitney test

All these tests are known loosely at $t$ tests, and they are calculated using the same Excel instructions up to the 'function argument' dialogue box. Wilcoxon is the related non-parametric test. Excel will not calculate the related parametric $t$-test. The unrelated $t$-test is parametric, and the Mann–Whitney test in the non-parametric version of the unrelated $t$-test.

For the Wilcoxon test, Table E.7 is used to illustrate some data based on Table 8.22. In the main dialogue boxes labelled home, insert, page layout, formulas, data, review, view > go to formulas > more functions > statistical > TTEST > the dialogue box 'function arguments'

**Table E.6** Presentation of data in Excel for a 2 × 2 contingency table. Based on Table 8.22.

|   | A | B | C | D |
|---|---|---|---|---|
| 1 | 6 | 11 | 9.06 | 7.93 |
| 2 | 10 | 3 | 6.93 | 6.06 |

**Table E.7** Data for the Wilcoxon test. Based on Table 8.22; percentage change in budgets.

|  | A | B |
|---|---|---|
| 1 | 8.56 | +5.0 |
| 2 | 25.1 | 0 |
| 3 | 11.53 | 0 |
| 4 | -7.18 | -1.1 |
| 5 | 38.83 | +15.3 |
| 6 | 3.45 | +2.5 |
| 7 | -4.56 | +1.0 |
| 8 | 29.98 | +8.3 |
| 9 | 23.92 | +6.2 |
| 10 | 11.34 | +4.8 |
| 11 | 1.87 | -3.0 |
| 12 | 7.68 | +4.5 |
| 13 | 1.83 | -2.8 |
| 14 | -5.46 | -7.2 |
| 15 | -0.96 | -1.0 |
| 16 | 2.71 | 0 |
| 17 | -0.03 | -0.1 |
| 18 | -10.86 | 0 |
| 19 | 6.71 | +6.6 |
| 20 | 8.22 | -0.2 |
| 21 | 25.66 | -2.2 |
| 22 | -9.53 | -5.2 |
| 23 | 8.31 | +8.2 |
| 24 | 5.13 | +2.0 |
| 25 | 3.65 | +1.2 |
| 26 | 13.95 | +3.2 |
| 27 | 17.44 | +10.0 |
| 28 | 4.85 | -3.0 |
| 29 | 0.68 | -2.2 |
| 30 | -2.02 | -2.9 |

appears with the text 'returns the probability associated with a student's $t$-test' > place the cursor in the cell labelled 'array 1' and the following text appears 'array 1 is the first set of data' > using the mouse, go to the spreadsheet main page and block in the cells A1–A30; this will give a reading in array 1 cell of A1: A30 > place the cursor in the cell labelled 'array 2' and the following text appears 'array 2 is the second set of data' > block in the cells B1–B30; this will give a reading in array 2 cell of B1: B30 > place the cursor in the cell labelled 'tails' and the following text appears 'specifies the number of distribution tails to return; one-tailed distribution = 1, two-tailed distribution = 2' > assuming a two-tailed test (see Chapter 8) type in '2' > place the cursor in the cell labelled 'type' and the following text appears 'is the kind of $t$-test: paired = 1, two-sample equal variance (homoscedastic) = 2, two-sample unequal variance = 3' > type in '1', the $p$ value will appear as $p = 0.002$

For the unrelated $t$ test, use the data in Table E.8 (based on Tables 8.19 and 8.20). Follow the same instructions as the Wilcoxon test, except block in the data cells A1–A17, and B1–B13 and in the last cell labelled 'type', insert 2, $p = 0.017$.

For the Mann–Whitney test, use the same data in Table E.8, and follow the same instructions as for the unrelated $t$-test, except in the last cell labelled 'type', insert 3, $p = 0.011$.

**Table E.8** Data for the unrelated *t* test and Mann–Whitney test. Based on Table 8.21.

| | A | B |
|---|---|---|
| 1 | 0 | +5.0 |
| 2 | −1.1 | 0 |
| 3 | +15.3 | +2.5 |
| 4 | +1.0 | +4.8 |
| 5 | +8.3 | −3.0 |
| 6 | +6.2 | −2.8 |
| 7 | +4.5 | −1.0 |
| 8 | −7.2 | −0.1 |
| 9 | 0 | −2.2 |
| 10 | 0 | −5.2 |
| 11 | +6.6 | −3.0 |
| 12 | −0.2 | −2.2 |
| 13 | +8.2 | −2.9 |
| 14 | +2.0 | |
| 15 | +1.2 | |
| 16 | +3.2 | |
| 17 | +10.0 | |

*Pearson's product moment correlation test—two routes both giving the same answer*

The data set in Table 8.26 is used to illustrate the test; represented in Table E.9.

Route one: in the main dialogue boxes labelled home, insert, page layout, formulas, data, review, view > go to formulas > more functions > statistical > Correl > the dialogue box 'function arguments' appears with the text 'returns the correlation coefficient between two sets of data' > place the cursor in the cell labelled 'array1' and the following text appears 'array 1 is cell range of values. The values should be number, names, arrays or references that contain numbers' > using the mouse go to the spreadsheet main page and block in the cells A1–A30; this will give a reading in the actual range cell of A1: A30 > place the cursor in the cell labelled 'array2' and the following text will appear 'array 2 is a second cell range of values. The values should be number, names, arrays or references that contain numbers' > block in the cells B1–B30; this will give a reading in the expected cell of B1:B30 > formula result will appear, with the correlation coefficient value given thus: formula result $p = 0.692$. Note this is the correlation coefficient, not the *p* value. To ascertain the *p* value, use tables for critical values of Pearson's *r*, where *r* must be equal to or more than the stated value to be significant.

Route 2: in the main dialogue boxes labelled home, insert, page layout, formulas, data, review, view > go to formulas > more functions > statistical > Pearson > the dialogue box 'function arguments' appears with the text 'returns the Pearson product moment correlation coefficient, r' > place the cursor in the cell labelled 'array1' and the following text appears 'array 1 is a set of independent values' > using the mouse go to the spreadsheet main page and block in the cells A1–A30; this will give a reading in the actual range cell of A1: A30 > place the cursor in the cell labelled 'array2' and the following text will appear 'array 2 is a set of dependent values' > block in the cells B1 to B30; this will give a reading in the expected cell of B1:B30 > formula result will appear, with the correlation coefficient value given thus:

**Table E.9** Data set for the Pearson correlation test. Based on Table 8.26.

| | A | B |
|---|---|---|
| 1 | +5.0 | +1.0 |
| 2 | 0 | +5.0 |
| 3 | 0 | 0 |
| 4 | −1.1 | +1.1 |
| 5 | +15.3 | +10.5 |
| 6 | +2.5 | 0 |
| 7 | +1.0 | 0 |
| 8 | +8.3 | +12.5 |
| 9 | +6.2 | +3.0 |
| 10 | +4.8 | +12.0 |
| 11 | −3.0 | 0 |
| 12 | +4.5 | +8.5 |
| 13 | −2.8 | 0 |
| 14 | −7.2 | −2.0 |
| 15 | −1.0 | 0 |
| 16 | 0 | −3.3 |
| 17 | −0.1 | +2.1 |
| 18 | 0 | +5.0 |
| 19 | +6.6 | −1.0 |
| 20 | −0.2 | 0 |
| 21 | −2.2 | +5.0 |
| 22 | −5.2 | 0 |
| 23 | +8.2 | +15.0 |
| 24 | +2.0 | +8.0 |
| 25 | +1.2 | +2.0 |
| 26 | +3.2 | +5.0 |
| 27 | +10.0 | +20.0 |
| 28 | −3.0 | +2.0 |
| 29 | −2.2 | +3.0 |
| 30 | −2.9 | 0 |

formula result $p = 0.692$. Note this is the correlation coefficient, not the $p$ value. To ascertain the $p$ value, use the table for critical values of Pearson's $r$ in Appendix M, where $r$ must be equal to or more than the stated value to be significant, and noting that the values are for the degrees of freedom, which are $n-2 = 30-2 = 28$.

# Appendix F: the standard normal distribution table

To determine the probability of a $z$ score, use the first column for the first decimal place, and the remaining columns for the second decimal place. Consider 1 standard deviation; the probability of scores being more than 1 standard deviation is 0.1587. The probability of scores being more than $\pm$ 1 standard deviation is twice that value $= 0.1587 \times 2 = 0.3174$. The probability of scores lying within $\pm$ 1 standard deviation $= 1.0 - 0.3174 = 0.6826$ or 68.26%.

| z | 0 | 1 | 2 | 3 | 4 | 5 | 6 | 7 | 8 | 9 |
|---|---|---|---|---|---|---|---|---|---|---|
| 0.0 | 0.5000 | 0.4960 | 0.4920 | 0.4880 | 0.4840 | 0.4801 | 0.4761 | 0.4721 | 0.4681 | 0.4641 |
| 0.1 | 0.4602 | 0.4562 | 0.4522 | 0.4483 | 0.4443 | 0.4404 | 0.4364 | 0.4325 | 0.4286 | 0.4287 |
| 0.2 | 0.4207 | 0.4168 | 0.4129 | 0.4090 | 0.4052 | 0.4013 | 0.3974 | 0.3936 | 0.3897 | 0.3859 |
| 0.3 | 0.3821 | 0.3783 | 0.3745 | 0.3707 | 0.3669 | 0.3632 | 0.3594 | 0.35557 | 0.3520 | 0.3483 |
| 0.4 | 0.3446 | 0.3409 | 0.3372 | 0.3336 | 0.3300 | 0.3264 | 0.3228 | 0.3192 | 0.3156 | 0.3121 |
| 0.5 | 0.3085 | 0.3050 | 0.3015 | 0.2981 | 0.2946 | 0.2912 | 0.2877 | 0.2843 | 0.2810 | 0.2776 |
| 0.6 | 0.2743 | 0.2709 | 0.2676 | 0.2643 | 0.2611 | 0.2578 | 0.2546 | 0.2514 | 0.2483 | 0.2451 |
| 0.7 | 0.2420 | 0.2389 | 0.2358 | 0.2327 | 0.2296 | 0.2266 | 0.2236 | 0.2206 | 0.2177 | 0.2148 |
| 0.8 | 0.2119 | 0.2090 | 0.2061 | 0.2033 | 0.2005 | 0.1977 | 0.1949 | 0.1922 | 0.1894 | 0.1867 |
| 0.9 | 0.1841 | 0.1814 | 0.1788 | 0.1762 | 0.1736 | 0.1711 | 0.1685 | 0.1660 | 0.1635 | 0.1611 |
| 1.0 | 0.1587 | 0.1562 | 0.1539 | 0.1515 | 0.1492 | 0.1469 | 0.1446 | 0.1423 | 0.1401 | 0.1379 |
| 1.1 | 0.1357 | 0.1335 | 1314 | 0.1294 | 0.1271 | 0.1251 | 0.1230 | 0.1210 | 0.1190 | 0.1170 |
| 1.2 | 0.1151 | 0.1131 | 0.1112 | 0.1093 | 0.1075 | 0.1056 | 0.1038 | 0.1020 | 0.1103 | 0.0985 |
| 1.3 | 0.0968 | 0.0951 | 0..0934 | 0.0918 | 0.0901 | 0.0885 | 0.0869 | 0.0853 | 0.838 | 0.0823 |
| 1.4 | 0.0808 | 0.0793 | 0.0778 | 0.0764 | 0.0749 | 0.0735 | 0.0721 | 0.0708 | 0.0694 | 0.0681 |
| 1.5 | 0.0668 | 0.0655 | 0.0643 | 0.0630 | 0.0618 | 0.0606 | 0.0594 | 0.0582 | 0.0571 | 0.0559 |
| 1.6 | 0.0548 | 0.0537 | 0.0526 | 0.0516 | 0.0505 | 0.0495 | 0.0485 | 0.0475 | 0.0465 | 0.0455 |
| 1.7 | 0.0446 | 0.0436 | 0.0427 | 0.0418 | 0.0409 | 0.0401 | 0.0392 | 0.0384 | 0.0375 | 0.0367 |
| 1.8 | 0.0359 | 0.0351 | 0.0344 | 0.0336 | 0.0329 | 0.0322 | 0.0314 | 0.0307 | 0.0301 | 0.0294 |
| 1.9 | 0.0287 | 0.0218 | 0.0274 | 0.0268 | 0.0262 | 0.0256 | 0.0250 | 0.0244 | 0.0239 | 0.0233 |
| 2.0 | 0.0228 | 0.0222 | 0.0217 | 0.0212 | 0.0207 | 0.0202 | 0.0197 | 0.0192 | 0.0188 | 0.0183 |
| 2.1 | 0.0179 | 0.0174 | 0.0170 | 0.0166 | 0.0162 | 0.0158 | 0.0154 | 0.0150 | 0.0146 | 0.0143 |
| 2.2 | 0.0139 | 0.0136 | 0.0132 | 0.0129 | 0.0125 | 0.0122 | 0.0119 | 0.0166 | 0.0133 | 0.0110 |
| 2.3 | 0.0107 | 0.0104 | 0.0102 | 0.0099 | 0.0096 | 0.0094 | 0.0091 | 0.0089 | 0.0087 | 0.0087 |
| 2.4 | 0.0082 | 0.0080 | 0.0078 | 0.0075 | 0.0073 | 0.0071 | 0.0069 | 0.0068 | 0.0066 | 0.0064 |
| 2.5 | 0.0062 | 0.0060 | 0.0059 | 0.0057 | 0.0055 | 0.0054 | 0.0052 | 0.0051 | 0.0049 | 0.0048 |
| 2.6 | 0.0047 | 0.0045 | 0.0045 | 0.0043 | 0.0041 | 0.0040 | 0.0039 | 0.0038 | 0.0037 | 0.0037 |
| 2.7 | 0.0035 | 0.0034 | 0.0033 | 0.0032 | 0.0031 | 0.0030 | 0.0029 | 0.0028 | 0.0027 | 0.0026 |
| 2.8 | 0.0026 | 0.0025 | 0.0024 | 0.0023 | 0.0023 | 0.0022 | 0.0021 | 0.0021 | 0.0020 | 0.0019 |
| 2.9 | 0.0019 | 0.0018 | 0.0018 | 0.0017 | 0.0016 | 0.0016 | 0.0015 | 0.0015 | 0.0014 | 0.0014 |
| 3.0 | 0.0013 | 0.0013 | 0.0013 | 0.0012 | 0.0012 | 0.0011 | 0.0011 | 0.0011 | 0.0010 | 0.0010 |
| 3.1 | 0.0010 | 0.0009 | 0.009 | 0.0009 | 0.0008 | 0.0008 | 0.0008 | 0.0008 | 0.0007 | 0.0007 |
| 3.2 | 0.0007 | 0.0007 | 0.0006 | 0.0006 | 0.0006 | 0.0006 | 0.0006 | 0.0005 | 0.0005 | 0.0005 |
| 3.3 | 0.0005 | 0.0005 | 0.0005 | 0.0004 | 0.0004 | 0.0004 | 0.0004 | 0.0004 | 0.0004 | 0.0003 |
| 3.4 | 0.0003 | 0.0003 | 0.0003 | 0.0003 | 0.0003 | 0.0003 | 0.0003 | 0.0003 | 0.0003 | 0.0002 |
| 3.5 | 0.0002 | 0.0002 | 0.0002 | 0.0002 | 0.0002 | 0.0002 | 0.0002 | 0.0002 | 0.0002 | 0.0002 |
| 3.6 | 0.0002 | 0.0002 | 0.0001 | 0.0001 | 0.0001 | 0.0001 | 0.0001 | 0.0001 | 0.0001 | 0.0001 |
| 3.7 | 0.0001 | 0.0001 | 0.0001 | 0.0001 | 0.0001 | 0.0001 | 0.0001 | 0.0001 | 0.0001 | 0.0001 |
| 3.8 | 0.0001 | 0.0001 | 0.0001 | 0.0001 | 0.0001 | 0.0001 | 0.0001 | 0.0001 | 0.0001 | 0.0001 |
| 3.9 | 0.0000 | 0000 | 0000 | 0000 | 0000 | 0000 | 0000 | 0000 | 0000 | 0000 |

# Appendix G: chi-square table

Critical value of the chi-square ($\chi^2$) distribution.
The calculated value of $\chi^2$ must be *larger* than or equal to the table value for significance.

| df | 0.10 | 0.05 | 0.025 | 0.01 | 0.001 |
|----|------|------|-------|------|-------|
| 1 | 2.706 | 3.841 | 5.024 | 6.635 | 10.828 |
| 2 | 4.605 | 5.991 | 7.378 | 9.210 | 13.816 |
| 3 | 6.251 | 7.815 | 9.348 | 11.345 | 16.266 |
| 4 | 7.779 | 9.488 | 11.143 | 13.277 | 18.467 |
| 5 | 9.236 | 11.070 | 12.833 | 15.086 | 20.515 |
| 6 | 10.645 | 12.592 | 14.449 | 16.812 | 22.458 |
| 7 | 12.017 | 14.067 | 16.013 | 18.475 | 24.322 |
| 8 | 13.362 | 15.507 | 17.535 | 20.090 | 26.125 |
| 9 | 14.684 | 16.919 | 19.023 | 21.666 | 27.877 |
| 10 | 15.987 | 18.307 | 20.483 | 23.209 | 29.588 |
| 11 | 17.275 | 19.675 | 21.920 | 24.725 | 31.264 |
| 12 | 18.549 | 21.026 | 23.337 | 26.217 | 32.910 |
| 13 | 19.812 | 22.362 | 24.736 | 27.688 | 34.528 |
| 14 | 21.064 | 23.685 | 26.119 | 29.141 | 36.123 |
| 15 | 22.307 | 24.996 | 27.488 | 30.578 | 37.697 |
| 16 | 23.542 | 26.296 | 28.845 | 32.000 | 39.252 |
| 17 | 24.769 | 27.587 | 30.191 | 33.409 | 40.790 |
| 18 | 25.989 | 28.869 | 31.526 | 34.805 | 42.312 |
| 19 | 27.204 | 30.144 | 32.852 | 36.191 | 43.820 |
| 20 | 28.412 | 31.410 | 34.170 | 37.566 | 45.315 |
| 21 | 29.615 | 32.671 | 35.479 | 38.932 | 46.797 |
| 22 | 30.813 | 33.924 | 36.781 | 40.289 | 48.268 |
| 23 | 32.007 | 35.172 | 38.076 | 41.638 | 49.728 |
| 24 | 33.196 | 36.415 | 39.364 | 42.980 | 51.179 |
| 25 | 34.382 | 37.652 | 40.646 | 44.314 | 52.620 |
| 26 | 35.563 | 38.885 | 41.923 | 45.642 | 54.052 |
| 27 | 36.741 | 40.113 | 43.195 | 46.963 | 55.476 |
| 28 | 37.916 | 41.337 | 44.461 | 48.278 | 56.892 |
| 29 | 39.087 | 42.557 | 45.722 | 49.588 | 58.301 |
| 30 | 40.256 | 43.773 | 46.979 | 50.892 | 59.703 |
| 31 | 41.422 | 44.985 | 48.232 | 52.191 | 61.098 |
| 32 | 42.585 | 46.194 | 49.480 | 53.486 | 62.487 |

# Appendix H: Mann–Whitney table, $p = 0.05$

Critical value of the Mann–Whitney $U$ statistic.

The calculated value of $U$ must be *smaller* than or equal to the table value for significance.

Two-tail $p$ or $\alpha = 0.05$

| n1 \ n2 | 1 | 2 | 3 | 4 | 5 | 6 | 7 | 8 | 9 | 10 | 11 | 12 | 13 | 14 | 15 | 16 | 17 | 18 | 19 | 20 |
|---|---|---|---|---|---|---|---|---|---|---|---|---|---|---|---|---|---|---|---|---|
| 1 | | | | | | | | | | | | | | | | | | | | |
| 2 | | | | | | | | 0 | 0 | 0 | 0 | 1 | 1 | 1 | 1 | 1 | 2 | 2 | 2 | 2 |
| 3 | | | | | 0 | 1 | 1 | 2 | 2 | 3 | 3 | 4 | 4 | 5 | 5 | 6 | 6 | 7 | 7 | 8 |
| 4 | | | | 0 | 1 | 2 | 3 | 4 | 4 | 5 | 6 | 7 | 8 | 9 | 10 | 11 | 11 | 12 | 13 | 14 |
| 5 | | | 0 | 1 | 2 | 3 | 5 | 6 | 7 | 8 | 9 | 11 | 12 | 13 | 14 | 15 | 17 | 18 | 19 | 20 |
| 6 | | | 1 | 2 | 3 | 5 | 6 | 8 | 10 | 11 | 13 | 14 | 16 | 17 | 19 | 21 | 22 | 24 | 25 | 27 |
| 7 | | | 1 | 3 | 5 | 6 | 8 | 10 | 12 | 14 | 16 | 18 | 20 | 22 | 24 | 26 | 28 | 30 | 32 | 34 |
| 8 | | 0 | 2 | 4 | 6 | 8 | 10 | 13 | 15 | 17 | 19 | 22 | 24 | 26 | 29 | 31 | 34 | 36 | 38 | 41 |
| 9 | | 0 | 2 | 4 | 7 | 10 | 12 | 15 | 17 | 21 | 23 | 26 | 28 | 31 | 34 | 37 | 39 | 42 | 45 | 48 |
| 10 | | 0 | 3 | 5 | 8 | 11 | 14 | 17 | 20 | 23 | 26 | 29 | 33 | 36 | 39 | 42 | 45 | 48 | 52 | 55 |
| 11 | | 0 | 3 | 6 | 9 | 13 | 16 | 19 | 23 | 26 | 30 | 33 | 37 | 40 | 44 | 47 | 51 | 55 | 58 | 62 |
| 12 | | 1 | 4 | 7 | 11 | 14 | 18 | 22 | 26 | 29 | 33 | 37 | 41 | 45 | 49 | 53 | 57 | 61 | 65 | 69 |
| 13 | | 1 | 4 | 8 | 12 | 16 | 20 | 24 | 28 | 33 | 37 | 41 | 45 | 50 | 54 | 59 | 63 | 67 | 72 | 76 |
| 14 | | 1 | 5 | 9 | 13 | 17 | 22 | 26 | 31 | 36 | 40 | 45 | 50 | 55 | 59 | 64 | 67 | 74 | 78 | 83 |
| 15 | | 1 | 5 | 10 | 14 | 19 | 24 | 29 | 34 | 39 | 44 | 49 | 54 | 59 | 64 | 70 | 75 | 80 | 85 | 90 |
| 16 | | 1 | 6 | 11 | 15 | 21 | 26 | 31 | 37 | 42 | 47 | 53 | 59 | 64 | 70 | 75 | 81 | 86 | 92 | 98 |
| 17 | | 2 | 6 | 11 | 17 | 22 | 28 | 34 | 39 | 45 | 51 | 57 | 63 | 67 | 75 | 81 | 87 | 93 | 99 | 105 |
| 18 | | 2 | 7 | 12 | 18 | 24 | 30 | 36 | 42 | 48 | 55 | 61 | 67 | 74 | 80 | 86 | 93 | 99 | 106 | 112 |
| 19 | | 2 | 7 | 13 | 19 | 25 | 32 | 38 | 45 | 52 | 58 | 65 | 72 | 78 | 85 | 92 | 99 | 106 | 113 | 119 |
| 20 | | 2 | 8 | 14 | 20 | 27 | 34 | 41 | 48 | 55 | 62 | 69 | 76 | 83 | 90 | 98 | 105 | 112 | 119 | 127 |

# Appendix I: Mann–Whitney table, $p = 0.01$

Critical value of the Mann–Whitney $U$ statistic.
   The calculated value of $U$ must be *smaller* than or equal to the table value for significance.
   Two-tail $p$ or $\alpha = 0.01$

|  | n1 | | | | | | | | | | | | | | | | | | | |
|---|---|---|---|---|---|---|---|---|---|---|---|---|---|---|---|---|---|---|---|---|
| | 1 | 2 | 3 | 4 | 5 | 6 | 7 | 8 | 9 | 10 | 11 | 12 | 13 | 14 | 15 | 16 | 17 | 18 | 19 | 20 |
| **n2** | | | | | | | | | | | | | | | | | | | | |
| 1 | | | | | | | | | | | | | | | | | | | | |
| 2 | | | | | | | | | | | | | | | | | | | 0 | 0 |
| 3 | | | | | | | | | 0 | 0 | 0 | 1 | 1 | 1 | 2 | 2 | 2 | 2 | 3 | 3 |
| 4 | | | | | | 0 | 0 | 1 | 1 | 2 | 2 | 3 | 3 | 4 | 5 | 5 | 6 | 6 | 7 | 8 |
| 5 | | | | 0 | 1 | 1 | 2 | 3 | 4 | 5 | 6 | 7 | 7 | 8 | 9 | 10 | 11 | 12 | 13 |
| 6 | | | 0 | 1 | 2 | 3 | 4 | 5 | 6 | 7 | 9 | 10 | 11 | 12 | 13 | 15 | 16 | 17 | 18 |
| 7 | | | 0 | 1 | 3 | 4 | 6 | 7 | 9 | 10 | 12 | 13 | 15 | 16 | 18 | 19 | 21 | 22 | 24 |
| 8 | | | 1 | 2 | 4 | 6 | 7 | 9 | 11 | 13 | 15 | 17 | 18 | 20 | 22 | 24 | 26 | 28 | 30 |
| 9 | | 0 | 1 | 3 | 5 | 7 | 9 | 11 | 13 | 16 | 18 | 20 | 22 | 24 | 27 | 29 | 31 | 33 | 36 |
| 10 | | 0 | 2 | 4 | 6 | 9 | 11 | 13 | 16 | 18 | 21 | 24 | 26 | 29 | 31 | 34 | 37 | 39 | 42 |
| 11 | | 0 | 2 | 5 | 7 | 10 | 13 | 16 | 18 | 21 | 24 | 27 | 30 | 33 | 36 | 39 | 42 | 45 | 48 |
| 12 | | 1 | 3 | 6 | 9 | 12 | 15 | 18 | 21 | 24 | 27 | 31 | 34 | 37 | 41 | 44 | 47 | 51 | 54 |
| 13 | | 1 | 3 | 7 | 10 | 13 | 17 | 20 | 24 | 27 | 31 | 34 | 38 | 42 | 45 | 49 | 53 | 56 | 60 |
| 14 | | 1 | 4 | 7 | 11 | 15 | 18 | 22 | 26 | 30 | 34 | 38 | 42 | 46 | 50 | 54 | 58 | 63 | 67 |
| 15 | | 2 | 5 | 8 | 12 | 16 | 20 | 24 | 29 | 33 | 37 | 42 | 46 | 51 | 55 | 60 | 64 | 68 | 73 |
| 16 | | 2 | 5 | 9 | 13 | 18 | 22 | 27 | 31 | 36 | 41 | 45 | 50 | 55 | 60 | 65 | 70 | 74 | 79 |
| 17 | | 2 | 6 | 10 | 15 | 19 | 24 | 29 | 34 | 39 | 44 | 49 | 54 | 60 | 65 | 70 | 75 | 81 | 86 |
| 18 | | 2 | 5 | 11 | 16 | 21 | 26 | 31 | 37 | 42 | 47 | 53 | 58 | 64 | 70 | 75 | 81 | 87 | 92 |
| 19 | 0 | 3 | 7 | 12 | 17 | 22 | 28 | 33 | 39 | 45 | 51 | 56 | 63 | 69 | 74 | 81 | 87 | 93 | 99 |
| 20 | 0 | 3 | 8 | 13 | 18 | 24 | 30 | 36 | 42 | 48 | 54 | 60 | 67 | 73 | 79 | 86 | 92 | 99 | 105 |

# Appendix J: Wilcoxon table

Critical value of the Wilcoxon $t$ statistic
The calculated value of $t$ must be *smaller* than or equal to the table value for significance.
Level of significance two-tailed test

| n | 0.10 | 0.05 | 0.01 |
|---|---|---|---|
| 5 | 1 | | |
| 6 | 2 | 1 | |
| 7 | 4 | 2 | |
| 8 | 6 | 4 | 0 |
| 9 | 8 | 6 | 2 |
| 10 | 11 | 8 | 3 |
| 11 | 14 | 11 | 5 |
| 12 | 17 | 14 | 7 |
| 13 | 21 | 17 | 10 |
| 14 | 26 | 21 | 13 |
| 15 | 30 | 25 | 16 |
| 16 | 36 | 30 | 19 |
| 17 | 41 | 35 | 23 |
| 18 | 47 | 40 | 28 |
| 19 | 54 | 46 | 32 |
| 20 | 60 | 52 | 37 |
| 21 | 68 | 59 | 43 |
| 22 | 75 | 66 | 49 |
| 23 | 83 | 73 | 55 |
| 24 | 92 | 81 | 61 |
| 25 | 101 | 90 | 68 |
| 26 | 110 | 98 | 76 |
| 27 | 120 | 107 | 84 |
| 28 | 130 | 117 | 92 |
| 29 | 141 | 127 | 100 |
| 30 | 152 | 137 | 109 |
| 35 | 214 | 195 | 160 |
| 40 | 287 | 264 | 221 |
| 45 | 371 | 343 | 292 |
| 50 | 466 | 434 | 373 |
| n | 0.05 | 0.025 | 0.005 |

Level of significance one-tailed test

# Appendix K; related *t* test table

Critical value of *t* for the related *t* test.

The calculated value of *t* must be *more* than or equal to the table value for significance. Note: one-tail significance levels at the foot of the table.

Level of significance for the two-tailed test

| df | 0.10 | 0.05 | 0.02 | 0.01 |
|---|---|---|---|---|
| 1 | 6.314 | 12.706 | 31.821 | 63.657 |
| 2 | 2.920 | 4.303 | 6.965 | 9.925 |
| 3 | 2.353 | 3.182 | 4.541 | 5.841 |
| 4 | 2.132 | 2.776 | 3.747 | 4.604 |
| 5 | 2.015 | 2.571 | 3.365 | 4.032 |
| 6 | 1.943 | 2.447 | 3.143 | 3.707 |
| 7 | 1.894 | 2.365 | 2.998 | 3.499 |
| 8 | 1.860 | 2.306 | 2.896 | 3.355 |
| 9 | 1.833 | 2.262 | 2.821 | 3.250 |
| 10 | 1.812 | 2.228 | 2.764 | 3.169 |
| 11 | 1.796 | 2.201 | 2.718 | 3.106 |
| 12 | 1.782 | 2.179 | 2.681 | 3.055 |
| 13 | 1.771 | 2.160 | 2.650 | 3.012 |
| 14 | 1.761 | 2.145 | 2.624 | 2.977 |
| 15 | 1.753 | 2.131 | 2.602 | 2.947 |
| 16 | 1.746 | 2.120 | 2.583 | 2.921 |
| 17 | 1.740 | 2.110 | 2.567 | 2.898 |
| 18 | 1.734 | 2.101 | 2.552 | 2.878 |
| 19 | 1.729 | 2.093 | 2.539 | 2.861 |
| 20 | 1.725 | 2.086 | 2.528 | 2.845 |
| 21 | 1.721 | 2.080 | 2.518 | 2.831 |
| 22 | 1.717 | 2.074 | 2.508 | 2.819 |
| 23 | 1.714 | 2.069 | 2.500 | 2.807 |
| 24 | 1.711 | 2.064 | 2.492 | 2.797 |
| 25 | 1.708 | 2.060 | 2.485 | 2.787 |
| 26 | 1.706 | 2.056 | 2.479 | 2.779 |
| 27 | 1.703 | 2.052 | 2.473 | 2.771 |
| 28 | 1.701 | 2.048 | 2.467 | 2.763 |
| 29 | 1.699 | 2.045 | 2.462 | 2.756 |
| 30 | 1.697 | 2.042 | 2.457 | 2.750 |
| 40 | 1.684 | 2.021 | 2.423 | 2.704 |
| 50 | 1.676 | 2.009 | 2.403 | 2.678 |
| 100 | 1.660 | 1.984 | 2.364 | 2.626 |
| df | 0.05 | 0.025 | 0.01 | 0.005 |

Level of significance for the one-tailed test

# Appendix L: Spearman's *rho* table

Critical value of the Spearman's *rho*
  The calculated value of $\rho$ must be *more* than or equal to the table value for significance.
  Level of significance two-tailed test

| *n* | 0.10 | 0.05 | 0.01 |
|---|---|---|---|
| 5 | 0.900 | 1.000 | |
| 6 | 0.829 | 0.886 | 1.000 |
| 7 | 0.714 | 0.786 | 0.929 |
| 8 | 0.643 | 0.738 | 0.881 |
| 9 | 0.600 | 0.683 | 0.833 |
| 10 | 0.564 | 0.648 | 0.794 |
| 12 | 0.506 | 0.591 | 0.777 |
| 14 | 0.456 | 0.544 | 0.715 |
| 16 | 0.425 | 0.506 | 0.665 |
| 18 | 0.399 | 0.475 | 0.625 |
| 20 | 0.377 | 0.450 | 0.591 |
| 22 | 0.359 | 0.428 | 0.562 |
| 24 | 0.343 | 0.409 | 0.537 |
| 26 | 0.329 | 0.392 | 0.515 |
| 28 | 0.317 | 0.377 | 0.496 |
| 30 | 0.306 | 0.364 | 0.478 |
| *n* | 0.05 | 0.025 | 0.005 |

Level of significance one-tailed test

# Appendix M: Pearson's *r* table

Critical value of the Pearson's *r*
  The calculated value of *r* must be *more* than or equal to the table value for significance.
  Level of significance two-tailed test

| *n*-2 | 0.10 | 0.05 | 0.01 |
|---|---|---|---|
| 2 | 0.900 | 0.950 | 0.9900 |
| 3 | 0.805 | 0.878 | 0.957 |
| 4 | 0.729 | 0.811 | 0.9172 |
| 5 | 0.669 | 0.754 | 0.875 |
| 6 | 0.621 | 0.707 | 0.834 |
| 7 | 0.582 | 0.666 | 0.798 |
| 8 | 0.549 | 0.632 | 0.765 |
| 9 | 0.521 | 0.602 | 0.735 |
| 10 | 0.497 | 0.576 | 0.708 |
| 11 | 0.476 | 0.553 | 0.684 |
| 12 | 0.457 | 0.532 | 0.661 |
| 13 | 0.441 | 0.514 | 0.641 |
| 14 | 0.426 | 0.497 | 0.623 |
| 15 | 0.412 | 0.482 | 0.606 |
| 16 | 0.400 | 0.468 | 0.590 |
| 17 | 0.389 | 0.456 | 0.575 |
| 18 | 0.378 | 0.444 | 0.561 |
| 19 | 0.369 | 0.433 | 0.549 |
| 20 | 0.360 | 0.423 | 0.537 |
| 25 | 0.323 | 0.381 | 0.487 |
| 30 | 0.296 | 0.349 | 0.449 |
| 35 | 0.275 | 0.325 | 0.418 |
| 40 | 0.257 | 0.304 | 0.393 |
| 45 | 0.243 | 0.288 | 0.372 |
| 50 | 0.231 | 0.273 | 0.354 |
| 60 | 0.211 | 0.250 | 0.325 |
| 70 | 0.195 | 0.232 | 0.302 |
| 80 | 0.183 | 0.217 | 0.283 |
| 90 | 0.173 | 0.205 | 0.267 |
| 100 | 0.164 | 0.195 | 0.254 |
| *n*-2 | 0.05 | 0.025 | 0.005 |

Level of significance one-tailed test

# Appendix N: *F* distribution

Critical values of the *F* distribution.
   The calculated value of $U$ must be larger than or equal to the table value for significance.
   $p$ or $\alpha = 0.05$

| Var1 | 1 | 2 | 3 | 4 | 5 | 6 | 7 | 8 | 9 | 10 | 12 | 15 | 20 | 24 | 30 |
|---|---|---|---|---|---|---|---|---|---|---|---|---|---|---|---|
| Var2 | | | | | | | | | | | | | | | |
| 1 | | | | | | | | | | | | | | | |
| 2 | 18.5 | 19.0 | 19.1 | 19.2 | 19.3 | 19.3 | 19.3 | 19.3 | 19.3 | 19.4 | 19.4 | 19.4 | 19.4 | 19.4 | 19.4 |
| 3 | 10.1 | 9.55 | 9.28 | 9.12 | 9.01 | 8.94 | 8.89 | 8.85 | 8.81 | 8.79 | 8.74 | 8.70 | 8.66 | 8.64 | 8.62 |
| 4 | 7.71 | 6.94 | 6.59 | 6.39 | 6.26 | 6.16 | 6.09 | 6.04 | 6.00 | 5.96 | 5.91 | 5.86 | 5.80 | 5.77 | 5.75 |
| 5 | 6.61 | 5.79 | 5.41 | 5.19 | 5.05 | 4.95 | 4.88 | 4.82 | 477 | 4.74 | 4.68 | 4.62 | 4.56 | 4.53 | 4.50 |
| 6 | 5.99 | 5.14 | 4.76 | 4.53 | 4.39 | 4.28 | 4.21 | 4.15 | 4.10 | 4.06 | 4.00 | 3.94 | 3.87 | 3.84 | 3.81 |
| 7 | 5.59 | 4.74 | 4.35 | 4.12 | 3.97 | 3.87 | 3.79 | 3.73 | 3.68 | 3.64 | 3.57 | 3.51 | 3.44 | 3.41 | 3.38 |
| 8 | 5.32 | 4.46 | 4.07 | 3.84 | 3.69 | 3.58 | 3.50 | 3.44 | 3.39 | 3.35 | 3.28 | 3.22 | 3.15 | 3.12 | 3.08 |
| 9 | 5.12 | 4.26 | 3.86 | 3.63 | 3.48 | 3.37 | 3.29 | 3.23 | 3.18 | 3.14 | 3.07 | 3.01 | 2.94 | 2.90 | 2.86 |
| 10 | 4.96 | 4.10 | 3.71 | 3.48 | 3.33 | 3.22 | 3.14 | 3.07 | 3.02 | 2.98 | 2.91 | 2.84 | 2.77 | 2.74 | 2.70 |
| 11 | 4.84 | 3.98 | 3.59 | 3.36 | 3.20 | 3.09 | 3.01 | 2.95 | 2.90 | 2.85 | 2.79 | 2.72 | 2.65 | 2.61 | 2.57 |
| 12 | 4.75 | 3.89 | 3.49 | 3.26 | 3.11 | 3.00 | 2.91 | 2.85 | 2.80 | 2.75 | 2.69 | 2.62 | 2.54 | 2.51 | 2.47 |
| 13 | 4.67 | 3.81 | 3.41 | 3.18 | 3.03 | 2.92 | 2.83 | 2.77 | 2.71 | 2.67 | 2.60 | 2.53 | 2.46 | 2.42 | 2.38 |
| 14 | 4.60 | 3.74 | 3.34 | 3.11 | 2.96 | 2.85 | 2.76 | 2.70 | 2.65 | 2.60 | 2.53 | 2.46 | 2.39 | 2.35 | 2.31 |
| 15 | 4.54 | 3.68 | 3.29 | 3.06 | 2.90 | 2.79 | 2.71 | 2.64 | 2.59 | 2.54 | 2.48 | 2.40 | 2.33 | 2.29 | 2.25 |
| 16 | 4.49 | 3.63 | 3.24 | 3.01 | 2.85 | 2.74 | 2.66 | 2.59 | 2.54 | 2.49 | 2.42 | 2.35 | 2.28 | 2.24 | 2.19 |
| 17 | 4.45 | 3.59 | 3.20 | 2.96 | 2.81 | 2.70 | 2.61 | 2.55 | 2.49 | 2.45 | 2.38 | 2.31 | 2.23 | 2.19 | 2.15 |
| 18 | 4.41 | 3.55 | 3.16 | 2.93 | 2.77 | 2.66 | 2.58 | 2.51 | 2.46 | 2.41 | 2.34 | 2.27 | 2.19 | 2.15 | 2.11 |
| 19 | 4.38 | 3.52 | 3.13 | 2.90 | 2.74 | 2.63 | 2.54 | 2.48 | 2.42 | 2.38 | 2.31 | 2.23 | 2.16 | 2.11 | 2.07 |
| 20 | 4.35 | 3.49 | 3.10 | 2.87 | 2.71 | 2.60 | 2.51 | 2.45 | 2.39 | 2.35 | 2.28 | 2.20 | 2.12 | 2.08 | 2.04 |
| 21 | 4.32 | 3.47 | 307 | 2.84 | 2.68 | 2.57 | 2.49 | 2.42 | 2.37 | 2.32 | 2.25 | 2.18 | 2.10 | 2.05 | 2.01 |
| 22 | 4.30 | 3.44 | 3.05 | 2.82 | 2.66 | 2.55 | 2.46 | 2.40 | 2.34 | 2.30 | 2.23 | 2.15 | 2.07 | 2.03 | 1.98 |
| 23 | 4.28 | 2.42 | 3.03 | 2.80 | 2.64 | 2.53 | 2.44 | 2.37 | 2.32 | 2.27 | 2.20 | 2.13 | 2.05 | 2.01 | 1.96 |
| 24 | 4.26 | 3.40 | 3.01 | 2.78 | 2.62 | 2.51 | 2.42 | 2.36 | 2.30 | 2.25 | 2.18 | 2.11 | 2.03 | 1.98 | 1.94 |
| 25 | 4.24 | 3.39 | 2.99 | 2.76 | 2.60 | 2.49 | 2.40 | 2.34 | 2.28 | 2.24 | 2.16 | 2.09 | 2.01 | 1.96 | 1.92 |
| 26 | 4.23 | 3.37 | 2.98 | 2.74 | 2.59 | 2.47 | 2.39 | 2.32 | 2.27 | 2.22 | 2.15 | 2.07 | 1.99 | 1.95 | 1.90 |
| 27 | 4.21 | 3.35 | 2.96 | 2.73 | 2.57 | 2.46 | 2.37 | 2.31 | 2.25 | 2.20 | 2.13 | 2.06 | 1.97 | 1.93 | 1.88 |
| 28 | 4.20 | 3.34 | 2.95 | 2.71 | 2.56 | 2.45 | 2.36 | 2.29 | 2.24 | 2.19 | 2.12 | 2.04 | 1.96 | 1.91 | 1.87 |
| 29 | 4.18 | 3.33 | 2.93 | 2.70 | 2.55 | 2.43 | 2.35 | 2.28 | 2.22 | 2.18 | 2.10 | 2.03 | 1.91 | 1.90 | 1.85 |
| 30 | 4.17 | 3.32 | 2.92 | 2.69 | 2.53 | 2.42 | 2.33 | 2.27 | 2.21 | 2.16 | 2.09 | 2.01 | 1.93 | 1.89 | 1.84 |
| 40 | 4.08 | 3.23 | 2.84 | 2.61 | 2.45 | 2.34 | 2.25 | 2.18 | 2.12 | 2.08 | 2.00 | 1.92 | 1.84 | 1.79 | 1.74 |
| 60 | 4.00 | 3.15 | 2.76 | 2.53 | 2.37 | 2.25 | 2.17 | 2.10 | 2.04 | 1.99 | 1.92 | 1.84 | 1.75 | 1.70 | 1.65 |
| 120 | 3.92 | 3.07 | 2.68 | 2.45 | 2.29 | 2.18 | 2.09 | 2.02 | 1.96 | 1.91 | 1.83 | 1.75 | 1.66 | 1.61 | 1.55 |
| ∞ | 3.84 | 3.00 | 2.60 | 2.37 | 2.21 | 2.10 | 2.01 | 1.94 | 1.88 | 1.83 | 1.75 | 1.67 | 1.57 | 1.52 | 1.46 |

# Index

---